T0297636

CLINICAL RESEARCH COMPUTING

CLINICAL RESEARCH COMPUTING
A Practitioner's Handbook

PRAKASH NADKARNI

Research Professor
Department of Internal Medicine, Carver College of Medicine
University of Iowa, Iowa City, IA, United States

Amsterdam • Boston • Heidelberg • London • New York • Oxford
Paris • San Diego • San Francisco • Singapore • Sydney • Tokyo
Academic Press is an imprint of Elsevier

Academic Press is an imprint of Elsevier
125 London Wall, London EC2Y 5AS, UK
525 B Street, Suite 1800, San Diego, CA 92101-4495, USA
50 Hampshire Street, 5th Floor, Cambridge, MA 02139, USA
The Boulevard, Langford Lane, Kidlington, Oxford OX5 1GB, UK

British Library Cataloguing-in-Publication Data
A catalogue record for this book is available from the British Library

Library of Congress Cataloging-in-Publication Data
A catalog record for this book is available from the Library of Congress

ISBN: 978-0-12-803130-8

For information on all Academic Press publications
visit our website at https://www.elsevier.com/

Working together
to grow libraries in
developing countries

www.elsevier.com • www.bookaid.org

Publisher: Mica Haley
Acquisition Editor: Rafael Teixeira
Editorial Project Manager: Ana Garcia
Production Project Manager: Julia Haynes
Designer: Matt Limbert

Typeset by Thomson Digital

CONTENTS

FOREWORD

Computer software plays a ubiquitous role in our lives. Marc Goodman [1] points out that about 100 million lines of program code control all the components of a high-end automobile, which is effectively a computer with wheels. The code that supports clinical research is possibly even larger: the bulk of this code, however, is invisible to the computing scientists and technologists who support research and represents off-the-shelf tools, such as database engines, spreadsheets, or statistical software which must be repurposed by programming in the environments provided by the tool vendors. The repurposing code possibly constitutes less than 1% of the total code, but it is critical because it applies general-purpose tools to specific clinical research problems. You, the reader of this book, are responsible for that code, either directly or indirectly (by supervising or liaising with others who write it). This book is your guide in helping you to deliver the goods, by helping you understand the science behind the problems.

MOTIVATION

The motivation behind my writing this book is twofold:

1. The primary motivation is, of course, to cover specific themes that the person supporting clinical research computing must be familiar with. You should not believe, however, that this will suffice. While any field of expertise advances continually, knowledge in the area of software engineering becomes obsolete at a particularly rapid clip. In a famous article in IEEE Software, Philippe Krutchen [2] estimated its "half-life"—here, the time required for half its body of knowledge to become superseded—at not much more than five years. Krutchen's article was written in 1988, and there is reason to believe that the half-life has shortened since then.

    Therefore, you must seek opportunities to expand your knowledge where necessary. "Experts" who fail to keep their knowledge current are an embarrassment to any field, but they can do particular damage in clinical research computing because the results of clinical research are ultimately applied to saving people's lives or making their lives better.

2. A secondary motivation is to give a feel for the processes of clinical research, as well to convey a sense of the skillset, and mind-set, that it takes to thrive in the research field and, by extension, the clinical research-computing field. It is said that much of research consists of borrowing other people's good ideas (and if you are ethical, acknowledging them). Clinical research computing similarly involves, where possible, reusing wheels created by someone else.

    To reuse an idea, however, involves having a sufficiently wide base of knowledge to realize that such an idea exists, and therein lays the challenge. No single individual can possibly hope to master all of the emergent knowledge in biomedicine and computing. One must be suitably humble in the face of this Niagara, while at the same time having an inner confidence that one will be able to step up and teach oneself (or train formally in) specific unfamiliar areas when this becomes necessary.

Going beyond one's intellectual comfort zone should be regarded not with fear but with a sense of adventure. One of my favorite Albert Einstein stories relates to a visitor at Princeton's Institute for Advanced Study, who found the great man studying an advanced undergraduate textbook. The book concerned a branch of mathematics that had not yet been discovered when Einstein had won his Nobel Prize 34 years earlier, and Einstein was exploring it in the hope of being able to apply it to his work on a theory that unified all the fundamental forces of the universe. (Incidentally, Einstein did not succeed in this effort. His time ran out a few months later, when he died of a ruptured abdominal aortic aneurysm in April 1955.) This is the kind of attitude I hope to see in everyone who does science for a living.

CHOICE OF TERMS

In my title for this book, I've used the general term "computing" rather than the specific term "informatics." I do this for several reasons.

- To solve problems in the field of clinical research computing, one must borrow solutions originally devised in the related areas of computer science, informatics, and information technology. In principle, computer science appears to emphasize theory, information technology focuses mostly on practice, and informatics fits somewhere in between. However, sterile hair-splitting about the boundaries between these fields seems to me to be an activity suitable only for those with too much leisure. Abundant spare time is a luxury that you are highly unlikely to have in the field of clinical research computing. Your clinical colleagues will bring problems to you, and your job is to devise solutions and possibly evaluate their effectiveness once implemented. In other words, you've simply got to do what you've got to do, and you can't afford to worry about boundaries.
- The word "informatics" derives from "information," which is traditionally data *after* it has been processed (somewhat). Computer programs need data (input) as higher animals need oxygen: it is the reason for their existence. Even back in the 1960s, the acronym GIGO ("garbage in, garbage out") indicated the importance of getting the data right for a particular computing problem. Modern user interfaces, as well as modern secure computing, depend to a large extent on recovering from erroneous inputs and providing helpful messages to the user as to what the correct input should be, or preventing deliberately erroneous inputs from doing harm to the system. More than half of the code of large systems deals with error recovery/prevention [3] and safeguards against possible errors in both software and hardware.

   The boundaries between the definitions of data, information, and knowledge are similarly fuzzy. All of them are inputs of varying degrees of refinement. In principle, data is raw, crude input; information is somewhat less crude (eg, summarized and annotated data); and knowledge is highly refined, actionable insight that is obtained by interrelating and culling multiple units of information. However, the choice of term used to define a particular input often depends on various factors.

   - *The computing problem to be solved*: Many data-processing tasks involve progressive refinement of inputs in a pipeline fashion in an attempt to make them more useful. One program's knowledge becomes another program's data.

- *The perspective of the author defining the terms:* Thus, Larry English's definition of information [4]—"information is data that is defined, summarized, and described *so as to be usable and meaningful*" (italics added)—implies a degree of quality. By this definition, misinformation would not constitute information and would be an oxymoron. The practical challenge, of course, is that just as you rarely detect a lie told to you until much later, you can't usually identify misinformation until sometime after it has been received, and possibly processed further or acted upon.
  - Occasionally, *the snake oil that is being sold:* Computing professionals who claim to deal with "knowledge" somehow seem more impressive than those who merely deal with data. The abundance of researchers prone to habitually slinging hype-laden words, with concrete achievements falling considerably short of expectations, gave the field of Artificial Intelligence a bad name in the 1980s and early 1990s.

    Be that as it may, transforming raw data into something useful, which often requires deep knowledge of the problem domain, is not necessarily the purview of the informatics professional alone.
- Finally, the neutral term "computing" implies both theory and practice. One of Karl Marx's still-quotable maxims from *Das Kapital* is "Practice without theory is blind; theory without practice is sterile." This book will provide a mixture of *theory*: the currently accepted principles that apply to specific problem subdomains, and *practice*: the experience of practitioners (and of one particular veteran—myself, biased though that viewpoint may be) that have been documented in the literature.

THE TARGET AUDIENCE

This book is intended to serve as a roadmap and detailed travel guide for two broad categories of readers.
- You may have come into clinical research computing from a computing area (computer science or information technology) or even the applied mathematics fields. You have either been roped into providing clinical research support, or else you've moved to clinical research because that's where many of the interesting problems are, and possibly as important, where much of the money increasingly is. (NSF funding has been decimated over the last decade, but no matter where your elected lawmakers' opinions lie regarding evolution or global warming, they're all afraid of dying prematurely of some disease.)
- Alternatively, you may be a clinical researcher who wants to better understand what the computing folks you collaborate with (or supervise) are up to. You may have possibly self-acquired some computing skills because you needed to solve a problem or do your job better, and you may have got bitten by the computing bug. Computing is less "mathematical" than it might seem to be (if your definition of mathematics is very narrow). People who enjoy solving puzzles and are skilled at games such as chess, bridge, backgammon, and Scrabble often find that their talent translates to computing, which they may find enjoyable and fulfilling. In any case, if you manage a team that includes software developers, you owe it to yourself to understand at least a little of what they do.
- If you're under 35, you may be a graduate of one of the many bioinformatics or health/medical informatics programs intended specifically for the burgeoning computing needs of the biological or clinical sciences. I've found that much of the theory that students are taught

bears little connection to reality. This is in part because much of the software that supports clinical research has been proprietary so that there are very few exemplars that may be studied in depth, unlike, say, the source code of the Linux operating system. To some extent, the quality of the software engineering in healthcare, in terms of functionality and usability, has left something to be desired, and the rate of improvement has been slow and irregular. The open-source movement is a relative newcomer, though it has had the benefit of putting pressure on the commercial vendors to enhance their offerings.

Also, many problems in clinical research support are not addressable by computing alone: typically, technology is the easiest aspect of the problem. For example, the lessons in Gerald Weinberg's 1971 classic, "The Psychology of Computer Programming," [5] are still relevant 45 years later. Some of the hard-earned lessons described in the present book have been learned through personal experience and are topics in which you are unlikely to have taken a course.

When writing for such a diverse readership, I have to walk a fine line between insulting the intelligence of particular readers by stating what may seem obvious to them and confusing everyone else through insufficient overview. I've chosen to err on the former side. I have provided sufficient sign posts in the text that follows that will allow knowledgeable individuals to skip particular portions of prose.

SCOPE OF THE BOOK

Due to the width of the field, topic coverage is necessarily synoptic, with maybe special emphasis in a few areas where I have had first-hand experience and can draw on this to provide additional perspective. I haven't addressed certain topics simply because I don't feel that I can do justice to them. For example, in the near and remote future, I believe that robotics will be increasingly important in its applications to prosthetics and rehabilitation, and this subfield deserves a book in its own right. However, I currently lack practical experience in this area, and I don't want to do a half-baked job pretending to impart a smidgen of knowledge that will, in Alexander Pope's words, only suffice to make you dangerous.

> *A note on style:* **As in my previous book, I've chosen to refer to myself as "I," and the reader as "you," with the justification of Admiral Hyman Rickover (as quoted by Ben Zimmer [6]) that only three types of individual are entitled to use "we": "a head of a sovereign state, a schizophrenic, and a pregnant woman."**
>
> **I've also chosen to write in an informal style. My idol is the great science and science-fiction writer Isaac Asimov (1920–1992)—a polymath who also wrote mystery novels, books on history, Shakespeare, the Bible, and Jewish humor. Asimov showed that one could convey extremely difficult, abstruse ideas most effectively by writing directly and simply, whereas a formal style might impair the clear transfer of information—as lawyers and the drafters of fine-print automobile contracts and software license agreements well know. One should not confuse passive voice and opacity with knowledge.**

Finally, I use a bold font for asides—text that is not directly related to the main topic, but is hopefully informative—such as the prose here. Readers who are skimming the book rather than reading it in depth can skip past such text.

ACKNOWLEDGMENTS

This book is dedicated to the memory of my brother, Dr Ravi Nadkarni (1939–2015), who was my mentor, guide, and best friend. Due to the considerable difference in our ages, he also served as a second father. He exemplified all the traits that are needed to succeed in any applied research field: wide-ranging knowledge, boundless curiosity, a focused and disciplined approach to problem solving, all coupled with the utmost scientific and moral integrity, as well as infinite patience with everyone less able than he.

While originally trained as a metallurgical engineer, Ravi could have excelled in almost any field he chose. Indeed, he later taught himself about environmental engineering and healthcare to the extent that he served on expert committees. He mentored numerous people besides me, and his advice was leavened with wit and humor. Ravi passed away before he could serve as an alpha tester for this book (as he did for my previous one), but his influence is here throughout. I continue to miss him every day.

Several friends and colleagues have served as beta testers to ensure the digestibility of this book's content, notably Drs Kimberly Dukes and Todd Papke, both of the University of Iowa. (Kimberly is a cultural anthropologist, Todd a computer scientist and informatician.) Any residual faults in this book, however, should be blamed solely on me.

BIBLIOGRAPHY

[1] M. Goodman, Security vulnerabilities of automobiles, Future Crimes, Doubleday, New York, NY, 2015, p. 243.
[2] P. Krutchen, The biological half-life of software engineering ideas, IEEE Software 25 (5) (1988) 10–11.
[3] S. McConnell, Code Complete, 2nd ed., Microsoft Press, Redmond, WA, 2004.
[4] L. English, Improving Data Warehouse and Business Information Quality: Methods For Reducing Costs and Increasing Profits, Wiley Computer Publishing, New York, NY, 1999.
[5] G. Weinberg, The psychology of computer programming, Silver anniversary ed., Dorset House Publishing, New York, NY, 1998.
[6] B. Zimmer, On language: "We". The New York Times, 2010.

CHAPTER 1

An Introduction to Clinical Research Concepts

1.1 INTRODUCTION

The reader with a clinical research background may choose to skip this chapter entirely or merely skim through it rapidly: it is oriented primarily toward the reader with a computing or nonclinical background whose knowledge of biomedicine is modest. I don't pretend to teach you all about clinical research in this brief chapter: I provide references (and brief reviews) of recommended reading in Section 1.8. If you don't know anything of life sciences or biomedicine at all, you would be advised to put this book down and skim through at least the first of the books in that list, and then come back to this chapter.

Clinical research is often classified dichotomously into the following categories.

- *Therapeutic studies:* These seek to determine *whether* a therapy works in a disease condition and to *what extent* it is effective. To ascertain the latter, it is typically compared with an established alternative therapy, if one exists. These studies vary in scale, based on the number of human subjects studied and the design of the study.
- *Mechanistic studies* (from "mechanism"): These try to find out *how* a therapy (typically an established one) influences biological processes in the body, either as part of a therapeutic effect or an undesirable one.

1.2 THE LEVEL OF EVIDENCE HIERARCHY

In the world of scientific questions (as well as in the everyday world), a viewpoint may be advocated by some and criticized by others. Topics of controversy include cause(s) of a particular disease (eg, fibromyalgia, chronic fatigue syndrome) or the effectiveness of a potential therapy—new, established or traditional—in a condition. The practical challenge is how to weight (ie, prioritize) the published evidence.

The most widely used aid for this purpose is called the "hierarchy of evidence," originally developed by the Canadian Task Force on the Periodic Health Examination and subsequently adopted by the US Preventive Services Task Force [1]. I will summarize it here and then point out the problems using this tool alone. I introduce the hierarchy because your team will have to implement systems (from scratch or by leveraging existing software) that support research designs at almost every level within the hierarchy, except possibly the lowest.

In order of increasing priority, the hierarchy is listed as follows:

1. *The case report* (based on observation on a relative handful of subjects, with the sample size as small as 1): This is given the lowest priority because of the small sample size, the possibility that other factors that could influence the observation (Section 1.6.1) were not accounted for, and the risk of personal bias. Nonetheless, historically, most of the great discoveries in medicine began this way.

2. *The cross-sectional study*: Here, data is collected from a larger number of patients at a single point in time (eg, when they show up in the hospital, or when organs or tissues are studied at biopsy or autopsy). One gathers as much data as possible, but the patients are not followed up, either because of lack of resources, or because it is physically impossible to do so.

   The bulk manufacture of cigarettes by James Bonsack's invention of the cigarette-rolling machine in Virginia, 1880, greatly reduced their price and eventually led to mass consumption. By the 1930s, a dramatic rise in the number of patients presenting with (previously uncommon) lung cancer was observed. This rise was linked to cigarette smoking through detailed case histories as well as autopsies on deceased patients. It is readily possible to differentiate the lungs of a dweller of a polluted city from a rural dweller, and in the latter, the lungs of a heavy smoker from a nonsmoker, simply by observing their color.

   The cross-sectional study shows associations between two observed phenomena, but association does not alone prove cause and effect—there may be an unexplored factor responsible for this association. With respect to cigarettes, this point was emphasized by the tobacco industry's big-name hired guns, including Ronald A. Fisher, founder of modern statistical science, and later the Yale epidemiologist Alvan Feinstein.

3. *Case-control studies*: These are systematic observations from two large groups of people from the same population who differ in an outcome of interest, for example, those who have the disease (cases) versus those who do not but are otherwise similar (controls). Such studies do not always require flesh-and-blood patients: one could simply perform retrospective analyses of Electronic Health Record (EHR) data.

   The improvement of the case-control study over the cross-sectional study in that the researcher identifies other factors that might influence the outcome and ensures that the individual combinations of conditions (called *strata*, Section 1.6.2) occur with approximately the same frequency in both groups. The effects of the condition can thus be estimated in isolation (in theory). However, there may still be one or more unknown factors that account for the differences between the groups.

4. *Cohort studies (longitudinal studies)*: These are observational studies where a large number of subjects with some shared characteristic (eg, dwellers of a particular geographical location, people who smoke, women using oral contraceptives) are followed up over long periods of time. One then determines which subjects develop particular outcomes (lung cancer, blood-clotting disorders) and which do not. Some cohort studies, such as the Framingham Heart Study in Massachusetts, have been going on for more than 50 years and have contributed much to our present knowledge of disease.

   Ideally, both case-control studies and cohort studies are *prospective* (ie, subjects are followed forward in time). Sometimes, because of the scarcity of prospective cohorts, one of the comparison groups may be *retrospective* (eg, using archival data from other studies as controls). However, a statistics dictum, which I first heard from David Schoenfeld of Harvard/Mass General Hospital, goes "Unlike wine, data does not age well with time." Retrospective data tend to be less reliable for several reasons.

   a. The veracity of cause and effect relationships is harder to establish.

   b. In the case of nonlaboratory data, (eg, symptoms and clinical findings), you can rarely go back to ask patients about particular findings known to be important today, but were not recorded in the old data: patients may be deceased or impossible to locate and are unlikely to provide you with that information even if alive and located.

   c. Similarly, when consistency checks reveal obvious data errors, they are usually impossible to correct for the same reasons as earlier. No one remembers minute details of what happened years ago.

   d. Data on important (and recently discovered) biomarkers relevant to the condition are rarely available in old data: a rare exception is old biopsy specimens, whose DNA can sometimes be analyzed. The current emphasis on biospecimen repositories is an attempt to address this problem: well-preserved specimens (urine, plasma, DNA) could be studied decades later for novel biomarkers.

5. *Randomized controlled trials (RCTs)*: These studies, typically conducted by multidisciplinary teams, prospectively compare two therapies for the same condition, and are discussed in depth in the subsequent sections. Such studies are likely to have been designed rigorously. Some trials will yield definitive results, where one therapy is shown unambiguously to be better than another, while others are nondefinitive—that is, the data *suggests* effectiveness but falls short of "statistical significance" (Section 1.3.3). Trials with definitive results are given more weight.

6. *Systematic reviews and metaanalyses of RCTs*: A *systematic review* is a study of the published literature on a particular theme (eg, therapy for a condition, uses of a particular medication) that combines the evidence in order to arrive at conclusions. The conclusions are then described in a narrative form and can serve as guidelines for the practitioner. *Metaanalysis* (research about previous research) seeks to augment the systematic review process by employing formal statistical techniques, from which data from several RCTs, both definitive and nondefinitive, is pooled to achieve higher *statistical power* (a concept discussed in Section 1.3.1) and arrive at more definitive conclusions than the individual studies themselves.

1.2.1 Limitations of the evidence hierarchy

The evidence hierarchy is not sacred: Bigby's [2] early criticism asked whether "the emperor has no clothes." The concerns are as follows.

1. Petticrew and Roberts [3] point out that a well-designed observational (cross-sectional, case-control, or cohort) study may be more valuable than a methodologically unsound RCT. McQueen [4] emphasizes that the hierarchy only tells you *how* a study was designed—it does not tell you whether the investigators were asking the right questions, or whether the design was appropriate to the questions.

2. The hierarchy's pinnacle, review/metaanalysis, may be agenda driven and systematically biased. This is because the reviewer has the freedom to make a judgment call regarding the merit of an individual study, which may be included or omitted from the analysis on various grounds. This freedom can be abused to "cherry pick." Even in honest metaanalyses, the benefit of the therapy being evaluated may be overestimated due to the "*file-drawer*" effect. That is, there is a publication bias because clinical studies that fail to achieve the expected benefit do not get published and end up in file drawers. Borenstein et al., in their excellent *Introduction to Meta-analysis* [5], point out correctly that the file-drawer and cherry-picking issues affect traditional reviews as well.

   In particular, the drug and device manufacturers have been accused, when seeking approval, of selectively presenting only the positive studies to the regulatory authorities. The US National Library of Medicine, working with the Federal Drug Administration (FDA), has tried to address publication bias with the clinicaltrials.gov site. All studies must be registered on this site so that patients with serious conditions looking to enroll in research studies may find the information to do so. Further, after the study is complete, the basic results, negative or positive, must be submitted.

3. As an extension of (2), the system can be gamed to influence scientific opinion. "Medical ghostwriting" [6] has achieved notoriety: systematic reviews in prominent journals, and even textbook chapters, which were attributed to prominent clinician-researchers, turned out to have been written by unacknowledged professional writers paid by drug/device manufacturers. The bigwigs then put their names on the works in exchange for sizable fees. A Senate investigation headed by Iowa Senator Charles Grassley explored the extent of the issue, which was particularly widespread in psychiatry.

4. Certain methodologies, such as those based on *qualitative research,* would tend to be placed at the lower levels of the evidence hierarchy.

Applied qualitative research employs data-gathering techniques such as participant observation, one-on-one interviews, or focus groups. These techniques were originally developed in the context of social sciences, cultural anthropology, industrial ergonomics, and marketing: many practitioners who originally trained in those fields are now making valuable contributions to healthcare, notably in studies of how it is implemented, delivered, and understood. Qualitative research is appropriate in studying relatively unexplored aspects of a problem. Here, formation of testable hypotheses may sometimes be postponed until *after* the collection of data (grounded theory). After inspection of the data, which is typically textual or narrative, recurring ideas/concepts are tagged with codes that can then be grouped into categories from which further high-level categories (themes) may emerge in turn. At this point, one might go on to quantitative methods.

For example, before formulating a questionnaire or survey, the results of which would yield numbers that can be analyzed quantitatively, it is desirable to first study the intended target population without any preconceived notions in order to determine what specific issues related to the topic under study are relevant (or of interest) to that population, and only then formulate the questions that will be asked. To do so otherwise is like surveying the population of a third-world village for their opinions on postmodern literary discourse.

As I discuss later in the chapter on Big Data, data mining is a bit like grounded theory in that it primarily *suggests* hypotheses rather than seeks to *confirm* then. The difference is that in data mining, the sheer volume of data mandates that much of the initial identification of leads be done electronically rather than manually.

Such a categorization, however, is unreasonable. For certain questions—including those directly related to this book's main theme—qualitative methods are the *only* ones that can provide answers. For example, software is validated by testing whether, given the correct inputs, it will provide correct outputs. However, for software intended for patient monitoring or medical-error prevention, validation does not suffice. Because it has to be used by fallible, fatigue-prone *human beings*, if the software has poor usability, its introduction may have an effect opposite to that intended. Usability testing is largely qualitative.

Software usability must be evaluated by studying subjects who resemble the target users and are ideally drawn from the same population. One observes them as they use the software and sees what errors they make. The subjects

are often asked to talk aloud as they operate the software, and finally asked in depth about the difficulties they face, what features would make their lives easier, and so on. This testing is often prolonged: some large companies like Apple and Microsoft will often release free "Community Previews" of new software, or new versions of existing software, to solicit input from thousands of users over months to a year.

Outstanding software disasters in nonhealthcare areas, such as Microsoft Windows 8 or the initial version of Apple Maps, occur when the company deliberately chooses to ignore the copious negative feedback it receives. Outsized egos typically play a role: the senior managers or division chiefs responsible for the product insist that they know better than the users what is really needed. The ideological fanaticism is accompanied by political maneuvering: any underling who is less than a "true believer" in the leader's vision is kicked out of the manager's group, or even fired.

With much "validated" and "certified" medical software, usability has been tested inadequately so that usability problems have not been identified and fixed. The software then requires a needlessly formidable learning curve, and after deployment, users—rather than the system—get blamed for mistakes that occur when using the software.

To be fair, responding usefully to user errors in a way that educates the user in which corrective steps to take is not easy. It requires extensive observation of users so that the sequence of events/user actions that lead to particular errors can be determined. The designers of numerous software packages fail to make these observations, and their error diagnostics are consequently highly misleading. Packages such as the Oracle DBMS are notorious for providing cryptic error messages in which the root cause could be anything and everything except what the software actually complains about.

The themes of software evaluation and usability issues are expanded on in Section 1.7.

Petticrew and Roberts, cited earlier, therefore propose augmenting hierarchy information with a matrix that emphasizes whether the methodology employed was *appropriate* to the research question. Their brief paper is essential reading.

In the next section, I introduce basic statistics as applied to clinical research because statistics principles influence research designs. If you're already fluent in biostatistics, skip this section because I'll be preaching to the choir. If, however, your statistics knowledge doesn't significantly exceed what you learned in high school, read on.

1.3 A BIRD'S-EYE VIEW OF STATISTICS IN CLINICAL RESEARCH

The field of statistics as applied to the life sciences and clinical research is vast enough that, as a career path, it existed long before the term "informatics" was even coined. I won't pretend to do justice to it here: I'm merely introducing it because adequate knowledge of fundamental statistics, which I believe is as essential as knowing how to brush your teeth or tie your shoelaces, is less widespread than it should be. The sterile presentation of the subject in many high school texts may be partly to blame. By contrast, Darrell Huff's 1954 classic, *How to Lie with Statistics*, the top-selling statistics book of all time and still in print, is a delight: it sweetens the pill through laughter, with amusing cartoons by the noted science cartoonist Irving Geis. (It may be rather hard to locate in most libraries, which categorize it under "humor" rather than "statistics.")

If you still need motivation to advance your statistics knowledge, read the following section.

1.3.1 Teaching yourself statistics: why

As a clinical research or computing professional, there are several reasons to sharpen your statistics skills.

- You will collaborate closely with biostatisticians as part of the research team. Learning something about what they do is necessary for your collaboration to be more effective, the communication barriers shrinking greatly as a result. Further, learning about other people's fields is a way of respecting what they do. (In multidisciplinary teams, those who believe that anything outside their field is not worth knowing, and who consciously or unconsciously convey lack of respect for others' expertise, soon become despised. Such people shouldn't be doing research.)
- The level of statistics you may have learned at the typical undergraduate level is just sufficient to hang yourself with. For example, you can't seriously do "Big-Data Analytics"—one of this decade's informatics buzzwords—or data mining without a foundation in multivariate statistics. This subject is, unfortunately, rarely taught as a core undergraduate course except to statistics majors. This is a legacy of the fact that it is impractical to do multivariate statistics without computer assistance.

1.3.2 Teaching yourself statistics: how

The good news is that there is a difference between learning how to *apply* advanced statistical techniques versus learning their theoretical basis or writing your own statistics programs, just as you don't have to be able to fix a car in order to drive it. Multivariate statistics, for example, relies on the branch of mathematics called matrix algebra, but you don't even have to be able to spell the words "determinant" or "eigenvalue"—both words refer to matrix-algebra concepts—to employ the techniques productively and appropriately. Thanks to the user-friendliness of numerous microcomputer statistics packages, self-education is easier than ever before. With the help of such a package, you can learn in a few weeks what used to take an entire semester or two in the years before such packages were available.

A word of advice here: the commercial (and fairly pricey) package SAS, while powerful, has too steep a learning curve to learn as a first tool. (Yale's Academic Software Support Team has more support staff for SAS than for all other microcomputer packages combined.) Its user-hostility may have contributed to the mystique around statistics as an arcane field of expertise. Packages like SPSS, Stata, Statistica, and MiniTab do about 80–90% of what SAS does, but with less than 10% of the learning effort.

Undergraduate science students in years past had to learn Latin and classical Greek. The chemist/physicist (and Nobelist) Wilhelm Ostwald (1853–1932) decried this practice on the grounds that it created an unwarranted respect for historical authority, which good scientists should be free of. SAS is the statistics equivalent of Latin. While part of the initiation process (or hazing ritual, if you're feeling less charitable), it doesn't make the typical user of statistical techniques more competent any more than knowing Latin makes a priest more moral or nurturing of his flock.

1.3.2.1 Cost considerations

The user-friendly packages are not cheap—about \$700–1000 full commercial price. If you work in an academic setting, however, an academic license is about 10% of the commercial price. Further, if your institution has a site license, the cost of specific packages may be nominal or even zero. If you're not in academia and don't want to spend money while you're learning, there is a lot of freeware to experiment with (see http://statpages.org/javasta2.html). Be warned, however, that freeware packages aren't as powerful (or even as good as learning tools) as the commercial packages. Some commercial systems (eg, SYSTAT) offer free scaled-down versions as well.

If you want freeware that has much more muscle than SAS, consider the R programming language. R was originally an open-source effort at the University of Auckland, New Zealand, intended to replicate the functionality of the S programming language (developed at Bell Labs in the Ma Bell era), but has leapfrogged the original product. R has a significant learning curve and a command-line environment rather than a graphical user interface, but can be rewarding if you put in the effort to learn it. In the last two decades, R's community (who keep contributing code) has grown by leaps and bounds, and among professional statisticians, R has been eating SAS's lunch. If you are a statistical analyst or data miner working on the bleeding edge, you are likely to find a novel technique implemented in R long before it makes it into commercial software. Third-party add-ons like Rstudio also provide a friendlier integrated development environment: Rstudio is available both as freeware and as greatly beefed-up commercial software.

1.3.3 Hypothesis testing and statistical inference in clinical research

Almost all clinical research needs to be evaluated using techniques of statistical inference (eg, when comparing treated vs. untreated groups of subjects, or subjects given a new therapy vs. subjects given a standard one). Does the therapy being evaluated really make a difference or could the results be a random fluke? Statistical inference allows us to determine the *probability* that something is a fluke (or not).

Essentially, statistical inference starts with a pessimistic viewpoint, the so-called *null hypothesis*—that is, an assumption that there is *no difference* between the groups (eg, a new therapy is not useful). Mathematically, this turns out to be easier to deal with than the *alternative hypothesis*—that the groups are different and the therapy is efficacious. One then tries to test the null hypothesis. There are two types of errors that one can make in testing.

- *Falsely rejecting the null hypothesis*: For example, concluding that the therapy works when in reality it does not (backing a loser). This error is also called a *Type I error*. The probability of a Type I error is called *Alpha*.
- *Falsely rejecting the alternative hypothesis*: For example, concluding that the therapy is a dud when it is really not (missing a winner). This error is also called a *Type II error*. The probability of a Type II error is called *Beta*.

Certain fundamental statistical concepts are derived.

- The ubiquitous term "*p value*" is an alternative term for alpha. Thus, when we say that the results of a therapy were significant at p less than 0.05, we mean that a Type I error is less than 5% likely. (In other words, we are 95% sure that therapy really helps.) One can never be 100% sure, but the surer we get, the closer alpha gets to zero.
- The *Statistical Power* is the number (1-Beta). When we say that a given experiment has a power of 80% (or 0.80), we mean that the chance of a Type II error is less than (1–0.80) = 20%.

In biostatistics, the "magic value" of p has been set at 0.05, for reasons discussed shortly. That is, if you are not at least 95% sure, you are not sure. The magic value of power is 0.8: in large federally funded studies, the sponsor may insist on a number of subjects sufficient to achieve a power of 0.9. With a study design with less than 80% power, a negative result—the usual outcome unless the novel therapy is miraculous—is impossible to interpret. It could be either because the therapy is really ineffective, or because the power was insufficient to detect an effect. I discuss these terms in more detail in the next two sections.

1.3.4 Type I errors: the choice of the *p* value

The 0.05 number was chosen through a tradition begun by Ronald A. Fisher. Many numeric phenomena in nature follow a so-called "normal" or *Gaussian* distribution, also known as the *Bell Curve*. The phrase "Bell Curve" describes the bell shape of a graph, using a vast number of samples of the number of times individual values occur (ie, the frequency) versus the value.

The Bell Curve happens to have specific mathematical properties. The peak of the graph (ie, the commonest value or *mode*) also happens to be the average value (or *mean*), and also happens to be the midpoint of the data if sorted in ascending order of value (or *median*). Further, 95% of the data points falling within two standard deviations of the mean (actually, 1.96 standard deviations, but 2 is a nice round number); 5% of values (2.5% on each side) would fall beyond two standard deviations (Fig. 1.1). (If you've forgotten the definition of standard deviation, read the section "Basic examples" in the "Standard Deviation" page of Wikipedia [7].)

> The name "Gaussian" honors the great Carl Friedrich Gauss (1777–1855) who first studied it and reported on the bell curve in 1809. Gauss coined the term "normal distribution" in a mathematical sense: it does *not* mean "typical" or "usual." However, it had been discovered much earlier, in 1733, by Abraham de Moivre, a Huguenot (French Protestant) refugee fleeing from persecution who,

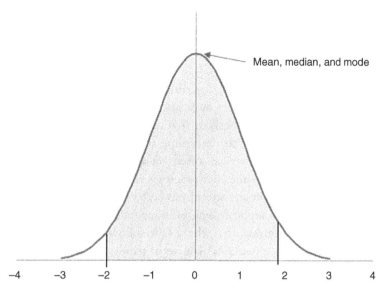

Figure 1.1 *The bell curve.* The curve is symmetrical about the mean, which is the peak of the curve, corresponding to zero on the *X*-axis. The values (ie, 95% of the area under the curve, as indicated by the shaded area) fall within 1.96 standard deviations on either side of the mean value.

> **after arriving in England, supported himself as a chess hustler in a London coffee house while doing math on the side.**

However, there is nothing sacred about 0.05. For proof of discovery of a new subatomic particle such as the Higgs Boson, a new chemical element, or telepathy, the threshold *p* value is much more rigorous—beyond four standard deviations, or 6.3×10^{-5}. Extraordinary claims require extraordinary evidence.

1.3.4.1 Multiple hypothesis correction

In certain circumstances, 0.05 is too optimistic. One of the standard techniques in "knowledge discovery" is to obtain a large dataset with dozens to thousands of variables. These may come, for example, from gene expression data, where thousands of gene products from tissue specimens are studied at the same time. You may compare gene 1's pattern with gene 2's, gene 1's with gene 3's, gene 2's with gene 3's, and so on, so that millions of comparisons are made. That is, you are testing *multiple hypotheses*.

In such situations, it can be shown through computer simulations with purely randomly generated data that, as the number of hypotheses increases, you will "strike gold" purely by chance, even though a subsequent experiment studying just those two variables will fail to show significant results. Therefore the *p* value must be adjusted downward. The simplest adjustment is called the *Bonferroni correction*, for which the threshold *p* value is set at 0.05 divided by the number of comparisons. Thus, for 1000 comparisons, the significant p value for any comparison would have to be 5×10^{-5} or less.

1.3.4.2 Data dredging

Data dredging is a derogatory term that describes an exercise in which researchers hell bent on publication repeatedly look for associations in a large dataset and publish after (1) claiming that only a single hypothesis or two was tested and (2) failing to repeat the analysis on an independent dataset to determine whether the findings are consistent. Califf [8] points out that a researcher can often get away with a faked claim. There are some giveaways, however, such as the methodology employed (it is almost always pure retrospective data analysis without any attempt to perform confirmatory experiments on animals, say), and the failure to replicate against independent datasets.

Data dredging has been known since the 1920s, following Karl Pearson's invention of the correlation coefficient, when a tidal wave of papers appeared correlating almost anything with anything else. (You can read more about this in the chapter "Post Hoc Rides Again," in Huff's book, cited earlier.) Tyler Vigen's hilarious book *Spurious Correlations* [9] illustrates correlations such as accidental swimming pool deaths in a given year versus the number of movies that Nicholas Cage appeared in that year.

A notorious example of data dredging, identified as such in a standard text [10], is a 1981 paper [11] by a Harvard group in the *New England Journal of Medicine* claiming an association between coffee and pancreatic cancer, which several subsequent studies by others failed to confirm. (Incidentally, *NEJM*, published by the Massachusetts Medical Society, is rated among the world's top medical journals. This paper provided ammunition for critics who assert that papers submitted from Harvard/Mass General get less vigorous scrutiny in *NEJM* than papers from "outsiders.") J.B. Rhine of Duke, who did experiments in the 1930s that claimed the existence of extrasensory perception, was also accused of dredging and file-drawer science (selective reporting, Section 1.2.1). No one succeeded in replicating Rhine's results.

1.3.5 Type II errors: power analysis and sample size

Power analysis is an application of statistics to determine the minimum sample size (eg, number of subjects) required to test a hypothesis while keeping alpha and beta to an acceptable minimum. The clearest explanation of power analysis that I've read is provided by Hill and Lewicki [12]. The necessary sample size depends on the following factors.

- The *type of the parameter used as the end point*: Numeric parameters (eg, change in blood pressure) require fewer samples then those that are *ranked* (eg, no/some/considerable/total pain relief), and these in turn require fewer than *binary* parameters (eg, survival/death, cure/no cure). For numeric variables, if the data is determined to fit a Gaussian frequency distribution, fewer samples are needed than if the data does not: an example of non-Gaussian data is income.
- *Intrinsic parameter variability between subjects/measurement error*: For most numeric data, this is measured by the standard deviation. The greater it is, the bigger the necessary sample size.

- *Experimental design*: With a "before/after" or *crossover* design (Section 1.6.3), where the same subject is given both treatments in succession, the requisite sample size is significantly less than if two separate groups of subjects were given different treatments, because variability between subjects does not have to be taken into account.
- The *expected differences* (ED) between treatments, expressed as a proportion (eg, the new therapy is expected to be 20% better than the old). Given the variation in response even for a single therapy across different subjects, this number should not be less the *minimum clinically important difference,* which may need to be estimated from the literature.

  For example, a study that measured self-reporting of pain intensity used a visual analog scale, where patients indicate their level of pain by making a mark on a 10-cm long line. A study by Lee et al. [13] found that, in patients with acute pain of less than 3 days' duration, a feeling of "adequate" pain control corresponded to a difference, on average, of 3 cm between the baseline pain level and the pain level after analgesia was administered and took effect.

  The ED is also important practically: the sample size goes up as the *inverse square* of the expected difference. That is, if you shrink the expected difference by a factor of 2, the sample size needs to be 4 times as large.
- The *alpha* and *beta* probabilities: The smaller these get (ie, the more rigorous the evaluation) the larger the required sample size. Sample size changes supralinearly with inverse beta. Thus, if the beta changes from 0.2 to 0.1 (ie, power changes from 80 to 90%), with a numeric variable the requisite sample size goes up by 33%. With alpha, if we move from approximately two standard deviations (0.05) to 2.5 standard deviations (0.01), the sample size goes up by 50%.

An excellent free program, *PS* (Power and Sample Size), by Dupont and Plummer of Vanderbilt University [14], lets you perform power analysis, as well as play with different values of the previously mentioned factors in "what-if" mode, to see the effect on sample size or power.

1.3.6 Interpreting negative results studies with adequate power and sample size

If the difference between a new therapy and an established one is marginal, the subjects may benefit from *both* therapies, but the comparison between them turns out to be nonsignificant. Assume that, for the standard value of alpha of 0.05 and beta at 0.8, and an expected difference (specified before the study began, obviously) of 10%, we find no significant difference. We can conclude that the *two therapies are less than 10% different from each other*, without any implied inference as to which is better.

I now return to this chapter's main theme. Clinical research studies can also be classified according to whether they concern *investigational* (novel) or *established* therapies. I discuss each in turn.

1.4 CLINICAL STUDIES OF INVESTIGATIONAL THERAPIES

Investigational therapies may include medications or medical devices. Studies of such therapies must be conducted under a regulatory/legal framework, enforced, in the USA, by the FDA. National frameworks get revised periodically, often in response to the publicity generated by adverse events of certain new therapies. Thus, the thalidomide disaster, where a number of babies were born in Europe/Britain missing one or more limbs after their pregnant mothers took this drug for morning sickness or as a sleeping pill. This resulted in regulations requiring extensive testing in animals for harmful effects on the fetus.

Clinical studies of investigational therapies are performed after extensive animal testing confirms efficacy and a level of safety appropriate to the condition being treated. They fall into three phases.

1.4.1 Phase I: early safety testing and dose determination

This phase, performed in about 10–20 volunteer subjects, who may not necessarily have the disease condition of interest—they may be healthy. In the case of investigational anticancer drugs, cancer patients who have not responded to any existing therapy might volunteer for Phase I studies. This phase has several objectives.

- *Evaluating the safety of the therapy*: If the therapy is a drug, one needs to determine the maximal dose of drug that can be tolerated by patients before severe side effects appear. The starting dose is estimated using animal-studies data, and then gradually higher doses are given (in different volunteers) and so forth. Eventually, one arrives at the highest dose compatible with an acceptable level of side effects. "Acceptability" depends on the condition that the drug is intended to be used for. Vomiting may be acceptable for an anticancer agent—many agents causing even severe vomiting are used in cancer therapy, with antivomiting agents given at the same time—but not in a drug intended to be used for headaches.
- *Pharmacokinetic* studies may also be performed for drugs. "Pharmacokinetics" is the study of what the body does to the drug (as opposed to *pharmacodynamics*—what the drug does to the body). Pharmacokinetic studies, which involve developing sensitive assays to measure the drug in body fluids, often tagging the drug molecule with a radioisotope to simplify detection, determine the following.
  - *To what extent the drug is absorbed when given by different routes*: orally, subcutaneously, intramuscularly, intravenously, by inhalation, or sometimes by uncommon routes such as in the nose, under the tongue, or applied to the skin. Exploration of different routes can partly be predicted by the molecular structure and physical properties of the drug—to be absorbed by the skin, for example, it helps if the drug does not have too high a molecular weight and is fat soluble. Some drugs are destroyed by stomach acid (eg, benzyl-penicillin, the original form of the antibiotic that was discovered by Alexander Fleming), and organic molecules with poor solubility in fat don't get absorbed well. (In the case of the painkiller morphine, the unabsorbed proportion that travels to the intestine can cause constipation—a side effect once used to treat diarrheas.)
  - *What parts of the body the drug goes to (distribution)*: Many drugs will not enter into cells, remaining only in the extracellular fluids: other drugs will not enter the brain, the microblood vessels of which actively keep many substances out. Substances such as the blood protein albumin or body fat can act as a depot or store for many drugs.
  - *What changes the drug undergoes in the body (biotransformation)*: Many drugs are first chemically modified—oxidized, reduced or broken down into smaller molecules. The liver is the most important organ for biotransformation, but other tissues can also participate. Some of the resulting molecules may themselves turn out to be active in the disease condition being investigated. Modification may be followed by *conjugation* or binding to other molecules such as glucuronic acid, sulfuric acid, and glutathione, which increase its

water solubility or neutralize it. Increased water solubility means decreased fat solubility: the conjugated molecule is less likely to be reabsorbed.

The ability of living creatures to change (and neutralize or get rid of) substances they have never encountered before is very ancient—going back to the bacteria. Some drugs may not undergo significant change at all in the relatively short time that they remain in the body. Other drugs, while absorbed, may be destroyed so rapidly in the liver (the first stop for digested food) that much less remains to work on the rest of the body: this is called the *first-pass* effect. (With morphine, 40–50% of an oral dose is inactivated in the liver.)

- *What routes it is excreted*: Drugs can be excreted through the urine (via the kidney), the intestine (either directly or most commonly via the bile), or in sweat and breath. Drugs excreted intact primarily by the kidney would tend to accumulate in patients with kidney failure.

The duration that the drug lingers in the body determines the frequency at which the drug's dose must be repeated. Sometimes, one can change the delivery mechanism of a drug to prolong its absorption, as in "sustained release" oral tablets, which consist of multiple layers that disintegrate successively. Similarly, pharmaceutical chemists may deliberately modify the drug's chemical structure to prevent destruction by stomach acid—for example, the orally administered penicillins—or increase fat solubility and consequently improve oral absorption. (Heroin, which is nothing but morphine with two acetyl groups attached, was originally synthesized to increase lipid solubility to better enter the body and thence the brain: in both the liver and the brain, it is converted back into morphine.) Modifying a drug so that it does not cross the brain at all has been used for morphine-like drugs that work on the intestine but have zero or minimal addictive potential.

Because of the small number of subjects tested in Phase I, only the most common side effects are discovered, and these are rarely the most dangerous. Similarly the dose determined is by no means definitive: in subsequent clinical practice, one may use much higher or lower doses.

In subsequent sections, please note that while the terms Phase II and III apply specifically to investigational drugs, the experimental designs used for Phase II/III studies are also (and in fact far more commonly) employed to study established therapies. This topic is discussed later in Section 1.5.

1.4.2 Phase II: the open therapeutic trial

The word "*open*" implies that the subjects know what drug they are being given, as opposed to a *blinded* trial, where patients do not know what drug they are receiving. The reasons for conducting blinded trials are discussed later in Section 1.6.4.

Just because a drug works in animal models is no guarantee of efficacy in humans. Phase II studies, the first step to measuring clinical efficacy, are performed in 20–40 subjects who have the disease being targeted. As in Phase I, subjects are vigilantly monitored for adverse effects, and one also gets a better idea of the dose range and drug administration schedule that would be employed in eventual practice.

Many drugs being evaluated for one purpose may be discovered to have an unexpected therapeutic benefit. Examples are chlorpromazine, a venerable antipsychotic (used for schizophrenia) that was originally being evaluated as an antihistamine for allergies. Lithium salts, which mitigate the manic phase of bipolar disorder and reduce relapses, were originally tested for gout, a painful joint (and sometimes kidney) condition caused by the accumulation or overproduction of the biological waste product uric acid.

The drug must be tested for an adequate period of time. Many drugs take time to show an observable beneficial effect: with antidepressants, for example, this is 6–8 weeks. This is because continued exposure to the drug results in adaptive changes within the body, such as a change in the quantity of specific molecules called *receptors*.

> **Receptors are molecules whose binding to other molecules (called *signaling molecules*) triggers some change or action. The vast majority of drugs act by binding to specific receptors. The action may occur in the same cell where the signal is released, or on neighboring cells, or in a faraway organ. There are thousands of different types of signaling molecules, with varied functions, and consequently thousands of different receptors. Signaling molecules released by nerves and brain tissues, which act on cells in the immediate vicinity (including themselves) are called *neurotransmitters*, while those that act on remote sites are termed *hormones*. An example of a hormone is insulin, which the body needs to process carbohydrates. Some substances can behave both as hormones and as neurotransmitters.**

1.4.3 Phase III: the comparative/controlled clinical trial

The objective of this phase is to determine how the new therapy compares with the current standard of care (eg, established drugs). The established therapy serves as a *control* (or a baseline for comparison). One must perform the comparison with numerous subjects with the disease condition, and ideally the studies are performed at more than one geographic site. One continues, of course, to monitor for adverse effects.

Ideally, a new therapy must have some comparative advantage over existing therapy. For medications, these might be one or more of the following: fewer adverse effects, less addiction potential (for painkillers) or suicide risk (if taken in overdose), longer duration of action, a more convenient route of administration, or better effectiveness in specific subsets of patients.

> **Many drugs may not meet any of these criteria. In fact, they may be so chemically similar to existing drugs that they are called "me-too" drugs. The cholesterol-lowering statins, and the various antianxiety drugs before them, are prime examples. The primary motivation behind their creation is to afford their creators a share of a lucrative market niche, but they sometimes have the unintended beneficial consequence of increasing competition between rival manufacturers in that niche and so driving prices down.**

As discussed in Section 1.5, comparative-effectiveness studies are hardly limited to Phase III studies. Two established therapies may be compared in subgroups of patients, or an established therapy may be investigated for a novel use based on a newly discovered pharmacological effect. An example of the latter was the investigation of the painkiller aspirin, beginning in the 1970s, to prevent heart attacks and strokes, after it was found to inhibit aggregation of platelets (the cells in the blood that help it to clot).

The area of comparative-effectiveness studies is a vast topic, and rather than introducing numerous subtopics here, I will address them separately from Section 1.6 onward.

1.4.4 Phase IV: postmarketing surveillance

After Phase III trials are completed, if the new therapy shows acceptable efficacy, the data on it is forwarded to the national regulatory agency (the FDA in the United States). If the agency is persuaded, approval is given and the therapy appears on the market. However, because of the modest number of subjects that have been screened, the full set of adverse effects will not have been discovered, simply because many effects are very rare. The antibiotic chloramphenicol, widely used in the 1950s and still kept in reserve for several infections, can cause fatal shutdown of production of bone marrow cells (aplastic anemia), but the incidence of this effect is estimated to be 1 in 40,000 patients.

Therefore monitoring of the therapy must never cease, and the period of postmarketing surveillance is sometimes referred to as Phase IV. Many a drug or device is either voluntarily pulled from the market or forcefully withdrawn by the regulatory agency, if the incidence of serious adverse effects is unacceptable for the condition where the therapy is being employed. Very rarely, however, a withdrawn therapy makes a modest comeback if it is discovered to have a novel property. This was the case with thalidomide—withdrawn in the 1960s, it made a mini-comeback in the 2000s as ancillary therapy for multiple myeloma, a form of cancer, after it was found to prevent angiogenesis (the production of small blood vessels: some cancer cells release angiogenetic substances to enable their own growth).

At one time, the primary responsibility for monitoring lay with the manufacturer who was supposed to report adverse effects to the national regulatory agencies. Such reports were received from practitioners or healthcare organizations, or from case reports and other studies in the medical literature. Because of the considerable duplication of effort involved at all levels, and the suspicion that not all manufacturers could be relied on for full disclosure, especially for "blockbuster drugs," surveillance responsibility shifted increasingly to the regulatory agencies. The surveillance was initially manual, but given the chronic understaffing at the FDA, the idea of automating such surveillance through innovative analysis of medical databases—notably EHRs, but also insurance claims databases—makes increasing sense. The FDA's Sentinel initiative [15] envisages a nationwide network for this purpose: pilot projects, notably "Mini-Sentinel" [16] are exploring the feasibility of doing this.

I now move to individual themes in research design.

1.5 CLINICAL STUDIES OF ESTABLISHED THERAPIES

"Established" therapies are those where a therapy has been officially approved by a regulatory authority for a condition, or where a therapy has been employed traditionally for a long time. (This would include alternative-medicine treatments such as acupuncture). The vast bulk of clinical research deals with established therapies, for several reasons.

1. Sometimes, based on the pharmacodynamics profile of a drug, we might want to test its effectiveness for a novel use. Because animal studies and Phase I studies are not needed, one would go directly into an open clinical trial and from there to a comparative-effectiveness study. While an institutional review board would be involved in deciding whether to approve the study, the national regulatory agency's involvement depends on the targeted use of the drug and its known adverse effects. (By the FDA's own admission [17], the cost and effort involved in submitting an application for regulatory approval of new clinical uses is onerous and may provide little benefit to a pharmaceutical sponsor for drugs that no longer have patent protection.) The FDA is in the process of simplifying the processes involved, at least in the context of anticancer drugs.

   A physician may sometimes employ a drug for an "off-label" use—one for which it is not officially approved by the regulatory agency. Sometimes, there may even be a rational basis for such therapy, and new therapies have been discovered this way. Consequently, such practices do not get scrutiny. However, if the off-label treatment was not done under the auspices of an IRB-approved study and something adverse happens to the patient, his family may get hold of a skilled lawyer who, after subpoenaing the medical records, then proceeds to crucify the physician in court: the hospital will then hang the physician out to dry. It is illegal for a manufacturer to promote off-label use. Pharmaceutical companies who encouraged physicians, through kickbacks, to employ certain drugs, have been fined millions of dollars. See the Wikipedia article [18] for a rogue's gallery of legal settlements.

2. In investigational-drug studies, the drug is given for a very limited time. Certain drugs, however, have different, or opposite, behavior in the short term and the long term, and these effects can only be measured through sustained observation. Thus, the drugs called beta-1-blockers worsen heart failure in the short term, but seem to reduce mortality in the long term. This "paradoxical," unexpected benefit was discovered relatively recently and demonstrates the importance of challenging commonly held beliefs on occasion.

   During heart failure, the body releases excess neurohormones (adrenaline) and neurotransmitters (noradrenaline) to make the heart beat more vigorously. Beta-1-blockers oppose these actions. In the short term, the pumping action weakens. In the long term, continued neurohormone release is like flogging a dying horse: damage to the heart muscle can result. Beta-1-blockers seem to protect against such damage.

3. Certain diseases are "silent killers" (notably, high blood pressure), with very few symptoms, and so the effects of therapy must be studied for many years to quantify the benefits of therapy on complications of the disease and mortality. In this regard, short-term studies often use a *surrogate* (or substitute) measure as the endpoint because it is relatively easy to observe or record, for example, suppression of an abnormal heart rhythm instead of long-term survival,

or cessation of a *febrile convulsion* (seen in susceptible children after the onset of fever) rather than quality of life. The parameter that really matters in the long term can only be quantified through extended observation.

At one time, doctors reflexively gave antiepileptic drugs for several years to any child who had a febrile convulsion, assuming that recurrent febrile convulsions led to progressive brain damage and drop in academic performance/IQ, along with an increased risk of chronic epilepsy and death from an unmanaged attack. When rigorous studies were done with untreated controls, it was found that (1) in untreated controls, the incidence of these conditions was much lower than previously supposed and (2) the drugs used to prevent epilepsy themselves had a whole battery of side effects, including irritability, lethargy, and sleep disturbances, leading often to the very drop in school performance they were supposed to prevent. In other words, to save one child from becoming epileptic, therapy created many more dullards, and its harm greatly outweighed its benefits. The American Academy of Pediatrics now recommends against long-term therapy [19]. Similarly, drugs that prevented heart-rhythm abnormality in the short term were found to increase mortality in the long term.

The issues with surrogate measures are similar to those related to "construct validity" in psychometrics—how accurately does a test that claims to detect scholastic aptitude, or a sociopathic personality, actually predict success in college, or habitual lawbreaking?

4. Investigational-therapy studies are necessarily limited in that the subjects selected to participate in those studies are a small subset of the patients in whom the drug will eventually be used. Children, for example, are often excluded from most studies, as are patients with severe concurrent disease, such as advanced kidney failure, so that the true risk of the therapy in these patients is not known. While one may estimate risk based on pharmacokinetic profile—for example, a drug excreted almost exclusively through the bile and intestine would possibly be safe in a patient with kidney disease—many of the adverse effects of long-term drug accumulation have been seen only when the drug was used in someone with some type of organ failure.

1.6 EXPERIMENTAL DESIGN OF COMPARATIVE-EFFECTIVENESS STUDIES

The first step in such studies is to identify all the factors that could bias the study outcome.

1.6.1 Factors influencing therapeutic response

Factors related to the therapy itself (eg, the drug, its dose, and route by which it is given) are only a few of the many that influence response. Other factors are:
*   The *severity* of the primary disease condition.
*   The *overall fitness* of the patient, and factors such as *age* and *gender*.
*   The presence of other *concurrent disease conditions* (eg, liver and kidney disease), which slow down the biotransformation or excretion, respectively, of several drugs.

- *Genetic factors* (of which only a relatively modest proportion have been identified).
- The *ancillary measures* that are used to treat the patient. Bed rest itself has a therapeutic effect in heart failure, and vigorous exercise mitigates depression.
- *Environmental factors,* of which again only a small proportion are known. Some foods may reduce the action of some drugs: the calcium in dairy products binds to the antibiotic tetracycline, reducing its absorption into the body. Other foods (eg, grapefruit juice) may prolong or increase the action of some other drugs by reducing their destruction in the liver. Concurrent medications may interact with the drug under study, either increasing its action or mitigating/canceling it. Some conditions, like depression and rheumatoid arthritis, often show spontaneous exacerbations or remissions due to unknown causes.
- The *placebo effect*—the patient's belief that the therapy is effective (Section 1.6.4).

These extraneous factors are also called *confounders* or *covariates*: both terms are borrowed from statistics. In a comparative-effectiveness study, these factors would obviously affect both the test therapy and the control therapy. The statistical design of the study, discussed shortly, is intended to minimize the effects of covariates.

An important determinant of experimental design is *whether the therapy potentially offers a reasonably long-lasting fix or cure,* or *whether the therapy only controls a chronic condition* as long as it continues to be administered. Novel surgical procedures or implanted prostheses or medical devices are examples of intended long-term solutions, as are treatments for most infectious diseases. Chronic conditions such as high blood pressure, diabetes, or major psychiatric illness are examples of the latter.

1.6.2 Separate patient groups: stratified randomization

For intended long-lasting treatments, one must use two separate groups of randomly selected patients, one given the new treatment and the other the standard treatment. Covariates are accounted for by a technique called *stratified randomization*. This is performed as follows.

- The research team first identifies the known factors that might influence response to therapy, such as those listed in the previous section.
- Recruited patients are then classified into different *strata* or groups, where a stratum is a particular combination of these known factors (eg, females below 40 with severe disease).
- Within a given stratum, one tries to assign roughly equal patients to each treatment through random selection.

The objective of stratification is to avoid accidental bias in assignment (eg, where all the severely ill patients get assigned to one therapy). There is, of course, the possibility that one has not accounted for unknown factors that might influence therapeutic outcome, but one has honestly done the best with the information that is known.

In both groups, some patients may experience minimal benefit while others experience dramatic benefit. In statistical terms, the standard deviation of the response may be large. Therefore average differences between the two treatment groups must be similarly large. Since, however, one can never know the average difference *a priori,* an enormous number of subjects must be studied to demonstrate that one therapy is better than the other.

1.6.3 Chronic conditions: crossover design

For chronic conditions where therapy only achieves temporary control, one can employ what is called a *crossover design.*
- An individual subject is first given either standard or investigational therapy. (Half may be started on therapy A, the other half on therapy B.)
- The therapy is maintained for a suitable period of time for the therapy to work, based on information gleaned from Phase II. Measurements that reflect clinical improvement (eg, fall in blood pressure or relief of symptoms) are performed regularly until the therapy has had enough time to act.
- After a while, the therapy is stopped. For medications, there is a *washout period*, during which the drug is allowed to be eliminated from the body: the washout period duration is informed by Phase I pharmacokinetic data.
- Then alternative treatment is given for the same duration of time.
- The effects of the two therapies are then compared.

The benefit of the crossover design is that one compensates for all the extraneous factors influencing therapeutic response (Section 1.6.1) because *each subject acts as their own control.* That is, assuming that experimental conditions did not change drastically between the times of the two therapies, any differences seen in response for a given patient are most likely to be due to differences in the therapies themselves. Further, the possibility that one drug may have sensitized, or partially nullified, the response to the other, is accounted for by varying the choice of starting drug, as stated earlier. Consequently, compared to the two-group design, the power of the study increases dramatically. That is, fewer subjects are needed to attain the desired power.

1.6.4 Placebo effects: double-blind designs

In many conditions, there is a large psychological component involved in how the patient reacts to treatment. The mere act of listening sympathetically, laying hands, and doing something that the patient believes will help the condition (either giving medication or performing a surgical procedure) can have a beneficial effect, even if the therapeutic measure is *known* to be inadequate or even useless. This is called the *placebo effect* from the Latin for "I please (you)." The placebo effect occurs because belief can have powerful physiological effects on the body through the release of neurotransmitters and hormones, and shut out even severe pain during times of intense stress.

> **At the Battle of Anzio (Italy, 1944), where US forces remained trapped on the beach for more than three months under incessant Axis shelling, the casualties were so numerous that supplies of morphine ran desperately low. The medic (and future famed researcher/bioethicist) Henry K. Beecher, who was forced to administer minimal morphine doses (diluted with saline), observed very dramatic relief in the soldiers who received them. Such doses would have been woefully inadequate in civilian injuries. Beecher reasoned that, unlike in civilian injuries, where an accident signals trouble, an injury at Anzio was a ticket out of the battle and an escape from highly probable death, and so the relief at being pulled out from the battlefront compensated for the inadequate dose.**

The placebo effect can also work in harmful ways. Patients who are allergic to a particular substance may develop a sudden reaction if given a placebo and told that they have been given the allergen. Other patients on a medication may spend much of their spare time reading on the Internet about its side effects, and then, sure enough, develop every subjective symptom in the book. This is a form of what has been called "cyber-chondriasis" (a fusion of "cyber" with "hypochondriasis").

The placebo effect is so widespread that studies of many therapies (eg, for psychiatric conditions, pain, high blood pressure, peptic ulcer, etc.) need to compensate for it. This is done through a "*double-blind*" design, where, after informed consent, (1) a subject is administered either the actual drug or a "dummy" medication, but not informed as to what she/he has received and (2) the clinician/nurse evaluating the patient is not given this information either. (Only pharmacists who are part of the research team, responsible for making both types of medication look alike, know what the patient has received.) Placebos may be used in a crossover design, so that the test treatment is contrasted with placebo, or as a third intervention in a study of two treatments, because it is possible that the observed benefits of *both* treatments may have a large placebo component.

Blinding the evaluator is necessary to avoid bias: the evaluator's attitudes about the drug (eg, that it is either a miracle drug or a deadly poison) can be subconsciously conveyed to the patient nonverbally. Blinding is not always possible (eg, for a highly invasive surgery). In cases where blinding is not possible, evaluation of the subject should be performed by an independent evaluator with no stake in the outcome.

Faked or "*sham*" surgery has been used in evaluation of minimally invasive procedures, where the risk imposed by the sham surgery is minimal. Its employment has been justified on the grounds that a truly independent evaluator, who draws salary support from the research effort, is usually a mythical figure. Many hallowed or experimental surgeries have been shown to be no better than placebo this way. Sham surgery is an ethical gray area, however, because you can't claim that the patient risk is always minimal. One way out (if the subsequent analysis shows that the treated group actually benefited) is to offer to the patient, during informed consent, the opportunity to be subsequently given the evaluated/effective treatment at no cost.

1.6.5 Pragmatic clinical trials

The highly controlled, artificial conditions under which the studies are performed are not necessarily the conditions that exist in real life (eg, patients may occasionally fail to take a dose, take the medication along with numerous others, and so on). A succinct, if technical way of saying this, is that these studies, if well-designed and properly conducted, may have high *internal validity* (ie, the conclusions that they reach are unbiased for the specific subjects studied), but low *generalizability* (external validity) to the population of patients as a whole.

Pragmatic clinical trials are a form of study design where the drug is employed exactly as it would be in practice, with average healthcare practitioners rather than specially trained researchers doing the evaluations as part of routine follow-up, and the EHR being employed for data collection for the most part, rather than a specially designed software system. This is a vast subject by itself: the following essential considerations apply.

- Because of the great diversity of patients, the differences between two therapies tend to be much more modest than in traditional research designs, and so a vast number of subjects need to be recruited for the study to have the requisite statistical power. (This is discussed in Section 1.3.1.)
- Pragmatic trials need to be overseen by an independent committee (*not* a drug sponsor) because they have a greater risk of bias, given the uncontrolled conditions in which they are performed, plus the sponsor's stake in the outcome.
- Compared to the traditional clinical trial, pragmatic trials need to be done "on the cheap" (calculated on a per-patient basis) because otherwise the research costs would be prohibitive. One aspect of this is that the data-entry burden on the practitioners, who are rarely (or minimally) compensated for their efforts, must be kept to a minimum. While Califf suggests [8] that systems supporting patient-entering of data may alleviate this burden, note that the requirement for minimal data entry remains. Patients forced to spend an excessive amount of time entering data may quit the trial in disgust, just as most of us hang up in the middle of an inordinately long telephone survey or poll.

1.7 EVALUATION OF MEDICAL SOFTWARE

In principle, the concerns that apply to medications, surgical techniques, and devices should also apply to medical software when the latter is employed for decision support in managing patients. After a few spectacular disasters, the FDA has become cautious and treats novel commercial medical software that is used directly in patient care as a medical device, subject to the same stringent safety checks (with the exception, of course, that animal/Phase I studies are not required).

Such caution is justifiable: many medical devices, including insulin pumps, electrocardiogram analyzers and cardiac pacemakers are controlled by software, and there have been disasters, such as when a software bug in a radiation-therapy device (Therac-25) ended up killing several patients with radiation overdose: the manufacturer shut down after the resultant lawsuits [20]. Among the numerous problems with this device (see the Wikipedia article [21] for more details) include the following.

1. *Usability*: numerous error conditions simply reported the noninformative diagnostic "MALFUNCTION: <error-number>". Users had no way to determine whether a particular error was innocuous or life threatening—the user manual did not explain or even address the error codes.
2. The users also had the option of overriding the diagnostic by pressing the "P" key to override the error and proceed. Like Pavlov's dogs, they soon became conditioned into doing this, because otherwise they couldn't get their work done.
3. Finally, the vendor had *never* tested the hardware/software combination until it was assembled at the hospital where it was used.

In the Windows/Macintosh world, such user-hostility as exemplified by 1–2 would be considered unacceptable in any software priced at more than $10.

> The stringent FDA regulations around medical software are the main reasons why the large-scale deployment of the IBM Watson software system for providing medical advice has been delayed. (Watson achieved fame on CBS's *Jeopardy!* television show. IBM subsequently announced that the Watson technology would be repurposed to assist medical diagnosis.) While IBM has publicly voiced their displeasure on the FDA's stand, I believe that the FDA is right. If IBM intends to make Watson a revenue source, they can't demand special treatment just because they have a big publicity machine that spoon-feeds lazy or credulous science journalists.

> The early examples of medical decision logic the IBM team provided in their numerous press releases included assertions such as "Given this patient's history, the probability of Lyme Disease is 73%." Such an approach is extremely naïve. First, any expert who could come up with such a suspiciously overprecise number would provide error bounds on the estimate (what is known in statistics as "95 percent confidence limits"): every estimate is vulnerable to sampling error. Second, to make reliable predictions, one must consider the rates' geographical, seasonal, or temporal variations. If busy and trusting doctors were led astray, such software could certainly do harm in perfect-storm circumstances, and IBM would share in legal culpability.

Interestingly, EHR systems are not treated this way, even though several dramatic implementation disasters occurred that disrupted workflow for weeks to months in the hospitals concerned. Adverse outcomes probably spiked as well, but since reporting of these is still largely voluntary and based on an honor system, there was little incentive to document them in depth and open the hospitals in question to legal action. At least one study, however, has recorded such an increase [22].

It is anybody's guess why this should be so. Historically, the first EHRs (many of which continue to be used) were developed in-house by academic research teams and the commercial systems then followed suit. I'm guessing that legislation mandating that academic systems retrospectively meet the onerous FDA certification requirements might be decried as posing undue hardship. Further, the commercial vendors might then protest about being selectively targeted, when at least some of the commercial systems are now superior to the in-house-developed ones.

Be that as it may, one informatician, Scot Silverstein of Drexel, believes that the deployment of at least some EHRs is essentially an experiment performed on human beings (patients and healthcare providers) without their consent [23]. I wouldn't go that far, but there is definitely scope for improvement. As with all large systems that cost tens of millions of dollars, the rate of software improvement is much less than in the medium-sized systems: changes are more difficult to make because of the large number of interlocking components, and also because competition is much less intense than for smaller packages. If you're a customer who's put up a vast amount of money up front, the vendor effectively owns you. It's very difficult to walk away, and so you may put up with inferior aspects of a system that you wouldn't think twice about discarding if it cost three orders of magnitude less.

1.8 FURTHER READING

1.8.1 Biomedical basics

Total beginners in the biomedical sciences would benefit from acquiring medical knowledge at least at a nursing student's level, so you should use nursing-student books as a first choice. The texts intended for medical students are heavy enough that your arms could get strengthened from hefting them regularly: so you should *refer* to them only after you are fluent in the basics. Try to read them cover to cover only if you have the time and inclination.

You start by learning about the normal body before learning what happens when things go wrong with it and how disease is treated. The multiauthor texts published by Lippincott, William, and Wilkins, such as *Anatomy and Physiology Made Extremely Easy!* [24] are a great place to start. These books are written by RNs, are aimed at nursing students, and are modestly priced, typically at less than $30. (They are also available as e-textbooks, a medium that I prefer myself.) From here, you learn pathophysiology (the process of disease), pharmacology (the study of medications), and then about clinical areas, such as the medical and surgical areas. If you are supporting basic-science researchers, a book such as *Molecular and Cell Biology for Dummies* is fairly decent. (Like the other "Dummies" books, the title is really a misnomer and just a come-on).

All this information takes time to master, but you should at least skim these books. (Done the right way, skimming can be a highly focused mental activity, because you are actively deciding what to ignore and what you retain.) This theme is expanded in next chapter.

1.8.2 Texts on clinical research

After you have acquired a smattering of the basics, you can start learning about clinical research itself. Reasonably good books that describe the clinical research itself are *Designing Clinical Research* (3rd edition) [25], by Hulley, Cummings, Browner, Grady, and Newman, and *Clinical and Translational Science*, a multiauthor work edited by Robertson and Williams [26].

Hulley et al. is more clearly instructive, more basic, and relatively brief, and should be read first. The second book is more advanced from the clinical perspective—clearly intended for a clinician—but necessarily provides only an introductory overview in other areas (biostatistics, computing). As tends to be typical of multiauthor works, individual chapters fluctuate in quality, and often introduce specialized terms without definitions, such as "confounding" (a subtle concept when employed in the statistical and data-interpretation sense), each author possibly assuming that another author has introduced the idea earlier. The book could have benefited from a glossary. Some chapters, however, such as Robert Califf's Chapter 2 [8], are superbly written.

BIBLIOGRAPHY

[1] B.I.Truman, C.K. Smith-Akin, A.R. Hinman, K.M. Gebbie, R. Brownson, L.F. Novick, R.S. Lawrence, et al. Developing the guide to community preventive services—overview and rationale. The task force on community preventive services, Am. J. Prev. Med. 18 (1 Suppl.) (2000) 18–26.

[2] M. Bigby, Challenges to the hierarchy of evidence: does the emperor have no clothes?, Arch. Dermatol. 137I 3 (2001) 345–346.

[3] M. Petticrew, H. Roberts, Evidence, hierarchies, and typologies: horses for courses, J. Epidemiol. Community Health 57 (7) (2003) 527–529.

[4] D. McQueen, The evidence debate, J. Epidemiol. Community Health 56 (2002) 83–84.

[5] M. Borenstein, L.V. Hedges, J.P.T. Higgins, H.R. Rothstein, Introduction to Meta-Analysis, Wiley, New York, NY, (2009).

[6] G. Schwitzer, Introducing a Three-Part Series on Medical Journal Ghostwriting. Health News Review, 2013.

[7] Wikipedia. Standard deviation. Available from: https://en.wikipedia.org/wiki/Standard_deviation, 2015.

[8] R.M. Califf, Clinical Trials, in: D. Robertson, G.H. Williams (Eds.), Clinical and Translational Science: Principles of Human Research, Academic Press, Los Angeles, CA, 2008.

[9] T. Vigen, Spurious Correlations, Hachette Books, New York, NY, (2015).

[10] D.L. Katz, J.G. Elmore, D.M.G. Wild, Jekel's Epidemiology, Biostatistics, Preventive Medicine, and Public Health, Saunders WB; Elsevier Health Sciences, Philadelphia, PA, (2014).

[11] B. MacMahon, S. Yen, D. Trichopoulos, K. Warren, G. Nardi, Coffee and cancer of the pancreas, N. Engl. J. Med. 304 (11) (1981) 630–633.

[12] T. Hill, P. Lewicki, Statistics: Methods and Applications, StatSoft, Tulsa, OK, (2007).

[13] J.S. Lee, E. Hobden, I.G. Stiell, G.A. Wells, Clinically important change in the visual analog scale after adequate pain control, Acad. Emerg. Med. 10 (10) (2003) 1128–1130.

[14] W. Dupont, W. Plummer, Power and sample size calculations: a review and computer program, Control Clin. Trials. 11 (1990) 116–128.

[15] Federal Drug Administration. FDA's sentinel initiative. Available from: http://www.fda.gov/Safety/FDAsSentinelInitiative/ucm2007250.htm, 2015.

[16] R. Platt, R.M. Carnahan, J.S. Brown, E. Chrischilles, L.H. Curtis, S. Hennessy, J.C. Nelson, et al. The U.S. Food and Drug Administration's Mini-Sentinel program: status and direction, Pharmacoepidemiol. Drug Saf. 21 (Suppl. 1) (2012) 1–8.

[17] Federal Drug Administration. Guidance for industry FDA approval of new cancer treatment uses for marketed drug and biological products. Available from: www.fda.gov/downloads/drugs/GuidanceComplianceRegulatoryInformation/ucm071657.pdf, 1998.

[18] Wikipedia. List of off-label promotion pharmaceutical settlements. Available from: https://en.wikipedia.org/wiki/List_of_off-label_promotion_pharmaceutical_settlements, 2015.

[19] American Academy of Pediatrics, Febrile seizures: clinical practice guideline for the long-term management of the child with simple febrile seizures, Pediatrics 121 (6) (2008) 1281–1284.

[20] N. Leveson, Safeware: system safety and computers, Addison-Wesley, Boston, MA, (1995).

[21] Wikipedia. Therac-25. Available from: https://en.wikipedia.org/wiki/Therac-25, 2015.

[22] A.X. Garg, N.K.J. Adhikari, H. McDonald, M.P. Rosas-Arellano, P.J. Devereaux, J. Beyene, J. Sam, et al. Effects of computerized clinical decision support systems on practitioner performance and patient outcomes, JAMA 293 (10) (2005) 1223–1238.

[23] Silverstein SI. In addition to nurses, doctors now air their alarm: Contra Costa County health doctors air complaints about county's new $45 million computer system. Available from: http://hcrenewal.blogspot.com/2012/09/in-addition-to-nurses-doctors-now-air.html, 2012.

[24] W.N. Scott, Anatomy and physiology made easy! Baltimore, MD: Lippincott William & Wilkins; 2015.

[25] S.B. Hulley, S.R. Cummings, W.S. Browner, D.G. Grady, T.B. Newman, Designing Clinical Research, 3rd ed., Lippincott Williams & Wilkins, Baltimore, MD, (2006).

[26] D. Robertson, G.H. Williams, Clinical and Translational Science: Principles of Human Research, Academic Press, Los Angeles, CA, (2008).

CHAPTER 2

Supporting Clinical Research Computing: Technological and Nontechnological Considerations

2.1 TECHNOLOGICAL ASPECTS: SOFTWARE DEVELOPMENT

In this chapter, I assume that you, the reader, have some control over your life in your day job in that you aren't simply taking orders from those above you, but also managing others. I'm guessing that you are responsible, not just for meeting with researchers and helping them crystallize the definitions of their problems, but also for having input in (or primary responsibility for) scoping out your team's participation in terms of the effort you will invest and how this effort will be compensated.

Three decades ago, much of the biomedical software used within a healthcare/research institution was custom built, either by the Health IT group or by an external development team under contract. The risks of purchasing large-system commercial software were commensurate with Russian roulette. Some systems ran on obsolescent hardware, with designs that were poster children for user hostility and very limited functionality. Small wonder that institutions like Latter-Day Saints' Hospital in Salt Lake City, Massachusetts General Hospital, the Veterans Administration, and Vanderbilt built their own EHR systems: some of these are still in use. Further, open-source systems in the biomedical field simply did not exist.

This situation is changing. Tools such as REDCap [1], created by Paul Harris's group at Vanderbilt University, greatly simplify the task of capturing research data. The i2b2 clinical-data-mart design created by Shawn Murphy and coworkers at Harvard has also become something of a de facto standard as a self-service tool for researchers who need to do sample-size estimations prior to applying for grants. Much commercial medical software, from the highly affordable to the high priced, is very functional and usable. The overall quality of the lower-end, specialized systems, as reflected in the satisfaction of their users, is considerably higher than for the high-end software: this is a direct consequence of the lower end of the market being much more competitive, a phenomenon seen in software in general.

Therefore the trend is increasingly toward buying a commercial system or using free software in the cases in which it fills a special niche. However, the need for custom software creation will never go away. Even purchased or open-source software must be integrated

Clinical Research Computing. http://dx.doi.org/10.1016/B978-0-12-803130-8.00002-6
Copyright © 2016 Elsevier Inc. All rights reserved.

with other systems, at least via data exchange, and this requires software development, albeit at a more basic level. Also, many EHRs, such as Epic™, provide data stores that allow complex queries, which run on machines separate from those running the application that healthcare providers interact with daily. Custom tools often need to be created to simplify the drudgery of performing standard tasks repeatedly against these stores.

2.1.1 Software-construction tasks: the development process

To those experienced with software development, the subsequent sections may seem like Software Engineering 101. However, I've observed that inappropriate approaches to software construction continue to be employed by teams that should know better. Read on if you're unfamiliar with development methodologies.

The traditional technique of creating software is called "waterfall." Systems analysts elicit a client's requirements in minute detail. They then create an equally detailed architecture document that specifies what will be done. Programmers then write code that rigidly adheres to that specification. The code is evaluated by testers to ensure that the software application conforms to the specification. Test failures are reported back and the code-test cycle continues until all tests pass.

The problems with this approach have been described extensively [2].

- The process is excessively rigid. Minor late-breaking changes, which are often necessary because of changes in legislation, financial climate, or business needs, are hard to accommodate because they require going back and overhauling the design specification.
- It is very difficult to accurately estimate project costs or anticipate technological risks because programmers are involved very late in the project. Competent/experienced programmers are the best judges of effort required, and they also know whether a particular requirement could be accommodated expeditiously with existing technology or would require leading-edge research to meet. A seemingly trivial feature request might actually be asking for the moon: if the clients were correctly warned about the delay and added cost entailed, or if an alternative approach achieving the same end-result were recommended, they would most likely drop that feature.
- Clients will find it hard to visualize the final product, so they are rarely able to specify requirements in depth through interviews alone. Consequently, a waterfall-designed product, after delivery, often turns out to be a camel (or worse, a goat) rather than the horse the customer expected. In his 1975 classic, *The Mythical Man-Month* [3], Fred Brooks acknowledged these issues, stating, "Plan to throw one (system) away – you will anyhow." The outcome is particularly disheartening if the client waited a long time, as is typical of waterfall development.

Alternative approaches are therefore almost always employed. I discuss a few of these, emphasizing that these approaches are typically combined.

2.1.1.1 Rapid functional prototyping

As stated earlier, in-depth interviews rarely suffice to get requirements right. A better way is to use the interviews as the basis for creating a *functional prototype,* an application that mimics the intended product, but which has been created relatively quickly, is greatly scaled down, and is far less robust than the final system will be. If the stakeholders are able to play with the prototype, their ideas gel much more effectively than if they were merely thinking in the abstract.

A prototype is called *functional* if users can actually enter and change data, and inspect the application's output, such as reports and charts—as opposed to a "werPoint prototype": a smoke-and-mirrors hustle in which someone shows you what the system *might* look like if only they had the money and people to build it. Microcomputer database packages are especially useful for creating functional prototypes. Using a graphical user interface similar to programs used for creating drawings, these prototypes can often be tweaked based on client suggestions in real time as the client peers over your shoulder.

Functional prototypes are of two kinds: *throwaway* prototypes and *evolutionary* prototypes. The former are created in a quick-and-dirty fashion as an aid to clarifying requirements and then discarded. Evolutionary prototypes are implemented with production-quality methods and tools: they take longer but are the foundation of the final product as they get increasingly refined. Karl Wiegers [4] points out that if you don't resist the temptation to move throwaway-prototype code into production, users will suffer the consequences. This temptation often arises under time pressure because a client is in a desperate hurry, and may be fooled into believing that software can be delivered almost as quickly as take-out pizza. Increasingly, however, many rapid-prototyping tools have become more capable over the years so that the code is robust enough to migrate to production if suitably modified. Many tools also allow large portions of the prototyping work to be migrated to a more scalable platform. Thus, microcomputer database schemas can often be ported with minimal effort to high-end, multiuser databases.

2.1.1.2 Early testing and continuous integration

Following the "prevention is preferable to cure" maxim, the modern trend is to incorporate basic tests for obvious incorrectness *into the code-development process itself*. Each programmer runs these "unit tests" before integrating one's code into the project's central code repository. Automated tools will even scan through program code and generate tests, which can then be modified as needed. Whenever one makes changes to the code, these tests are rerun to ensure that no test fails.

Individual programmers' code in the repository is combined (*integrated*), and testers then use software that records and plays back user actions (keystrokes and mouse clicks) to perform "integration tests." In *continuous integration*, some integration testing is overlapped with development. That is, the automated integration tests are run each time code is added to (or changed in) the repository, though, understandably, other tests can only be done on the putative final product.

2.1.1.3 Evolutionary delivery

As opposed to the "big bang" approach of waterfall, evolutionary delivery is based on delivering *increasing functionality* (starting with the most important) in stages, typically every month or two. Steve McConnell [5] points out several benefits of staged delivery.

- *Reduced risk*: The most important and critical functionality is tackled (and delivered) earlier. Therefore problems are more likely to be detected early. The client is in a better position to provide corrective feedback before too much damage is done. Besides, the client can *see* progress being made as the application evolves: status reports on how work is progressing become superfluous.
- *Reduced estimation error*: Estimates of work made at the start of the project may be off by a factor of 10 in either direction. As one actually starts doing the work, estimates become progressively more accurate. Staged delivery, in which work is started early after requirements gathering, allows more accurate and frequent recalibration.

Evolutionary delivery has a modest downside in that project management needs to be much tighter. Further, automation of testing becomes mandated because, with each release, tests must be rerun. In other words, to use evolutionary delivery effectively, your team has to elevate all aspects of its game.

2.1.1.4 "Agile" development

Prototyping, continuous integration, and evolutionary delivery are aspects of a family of methodologies called "Agile Development." Several good books exist on the subject, such as Stellman and Greene [6]. There are also blogs about "good" versus "bad" "agile" [7], which explore the pros and cons of agile development in detail. I won't go further than referring you to the fairly balanced Wikipedia article [8]. Most "agile" practices are now considered mainstream, though some—like "pair programming," where two programmers sit together and take turns writing code—turned out to be temporary fads.

I'll also reiterate that no matter what methodology you use, certain fundamentals don't go away. Thus, you owe it to the client and your team to try to get requirements in some depth first, rather than trying to create an application (even a prototype) straight away. McConnell's take on Brooks' dictum goes: "If you plan to throw one away, you'll end up throwing away two." [9].

Further, "agile" is not a *carte blanche* (as some naïve programmers believe) to create half-baked, disorganized documentation (à la Microsoft Office 2007 and later) that is almost useless to the average user, or not to create documentation at all. (Microsoft, as a near-monopolist in the word processor/spreadsheet world, gets away with it: you may not be able to.)

2.1.1.5 Use the simplest technology that meets requirements

In supporting clinical research, it is possible for the typical small team to be completely drowned in work if they use technical approaches that are inappropriately complex when compared to the task at hand. Many users, for example, want a multiuser database that will be used by a maximum of three concurrent users within the same institution. Using a microcomputer database framework to build the application takes about a 10th of the time (conservatively: real gains are typically greater) than building the equivalent web-based application would entail. Sure, it won't scale, but here, speed of development trumps scalability: the latter is a nonissue. *If* the client comes back later requesting something scalable (and pays you enough for the added development), you can cross this particular bridge.

2.2 NONTECHNICAL FACTORS: OVERVIEW

The most important resources in supporting clinical research are the people involved—you and your team. Studying the psychological and sociological factors that make for effective individuals and teams is, understandably, an active area of exploration in computing, and I'll summarize the essentials here.

Supporting the computing aspects of clinical research requires several skills and traits.
1. A very specific *attitude* that makes you a good match to the job: without it, you're a misfit.
2. *Technical and domain competence,* coupled with the desire to *stay* competent. The mere fact that you were knowledgeable 10 years ago means little today if you haven't kept up.
3. *General and nontechnical skills:* These include the ability to communicate your ideas effectively both orally and in writing, and people skills, notably the ability to negotiate and support your own team.
4. *Specific personality traits* that make you more likely to succeed.

Of these four, only the last is mostly innate; the others can be acquired through self-discipline and effort. I'll now elaborate on each topic.

2.3 ATTITUDE: SERVICE VERSUS RESEARCH

Clinical-research computing is different from some other computing/informatics fields in that, if you're in academia, a large part of your salary comes from supporting the efforts of other people—specifically clinical and basic-science researchers—rather than from teaching or doing your own research. In other words, just like many researchers who spend some time treating patients, you perform a *service* role. Many professional biostatisticians, computer scientists, and even mathematicians also operate this way, and not just in clinical research alone. Your success in service is measured by the extent to which you make *others* more productive and efficient.

Such research as you are able to do is typically applied and is a direct or indirect consequence of devising solutions to the problems that people bring to you. Such solutions have two aspects.
1. They may be *radically better* than the current solutions. Possibly the most famous example of this is the Fast Fourier Transform (FFT), an algorithm devised by James Cooley and John Tukey in 1965 [10].

   Through the invention of electronic circuits that implement it, the FFT has impacted fields as diverse as signal transmission (satellite, radio, and cable), electronic music, earthquake detection, and image processing. The motivation

for its creation was at a meeting of President Kennedy's Scientific Advisory Committee that was discussing the challenge of detecting nuclear-weapon tests in the Soviet Union by employing seismometers located outside the USSR. In 1984, it was discovered that an unpublished paper, written in Latin by Carl Friedrich Gauss (of the Gaussian distribution), had developed the same idea in 1822, using it to calculate the orbit of asteroids.

2. They may be more *general* or *generalizable*. The motivation for such solutions is "creative laziness." Smart (and busy) computing professionals who are paid only after a job has been successfully accomplished (rather than by the hour) encounter the same kind of problem repeatedly, or foresee the strong possibility of this happening. Instead of performing the same tedious series of steps each time, they work extra hard to implement general solutions that either semiautomate these tasks or better still, make the jobs doable by a less skilled underling or even a nonprogrammer. In either case, the solution frees up one's time to focus on the next interesting problem, rather than being buried in routine drudgery. Many open-source projects have begun this way.

Ben Franklin's maxim, "When you want something done in a hurry, you ask a busy person," needs to be slightly modified here. The first time, the job takes longer because the general solution is harder to devise. Subsequent jobs, however, get done very quickly. The enlightened manager therefore cuts one's creative people some slack: creativity needs this initial time investment in order for more optimal solutions to be devised. Additionally, talented people who are not given the freedom to devise smarter solutions soon quit.

2.3.1 Balancing act

Most computing support teams are modestly sized; unlike in large organizations, you do not have the luxury of creating a pure-research group. Herein lays a challenge.

Unless you and your people have some time blocked off to think and reflect (as well as to keep up with advances in the field), you will fall behind intellectually. The corollary to Franklin's maxim is: "The busy person gets dumped on." At that point, your environment resembles labor organizer Saul Alinsky's definition of Hell as "a place where one does the same thing over and over." [11].

However, if you take someone's money, you are obligated to provide value in return. If you are in a service job and you spend almost all your supported time in research, you may keep your clients waiting indefinitely. When (not *if*) their patience runs out, they will either hire someone else to meet their needs, try to get you fired or laid off, or both.

In some organizations, such as 3M, Google, Facebook, and Amazon, allowing employees to block off 15–20% of their hours to work on their own problems is standard policy. Academic centers, however, do not have any similar official policy for nonfaculty, who are often expected to do this on their own time. And even though faculty members are expected to stay on the leading edge, the part of their time that is supported by service must be dedicated to service.

There is a way out, assuming here that you don't work in an intellectual sweatshop—some of these exist in academia too—and that most of your collaborators and clients are reasonable. When you first join the group, especially if you're not on faculty, you have to earn the right to spare time. You do so by accumulating a reservoir of goodwill through reliable and quality delivery for at least 6 months to a year. Once your colleagues trust your competence and judgment, you are then in a position to negotiate, both for yourself and for your team members, for some time to be blocked off for self-development and devising of better solutions. If your collaborators aren't totally myopic or malicious, they will see the long-term benefits that can accrue from this modest investment. Notably, you and your team will be able to deliver better solutions with fewer personnel requirements than other teams, because you have been given permission to work smarter.

2.4 TECHNICAL SKILLS

At one time, "medical informatics" in many hospital settings simply meant acting as a liaison between clinicians and software developers. Such individuals were typically clinicians who had done a fellowship rather than a degree program, or else they were simply more turned on by computers than their brethren. They normally played the informatician role on the side, being primarily clinicians or researchers. Any computing skills that they possessed were almost entirely self-taught.

A pure liaison role made sense in the days when most clinicians were computer-phobic. Today, however, the typical clinical resident or junior-to-mid-level faculty member has grown up using and playing with computers—and medical students are increasingly ex-computer science, engineering, or even math majors. Clinicians can increasingly communicate directly with computing professionals, and more important, the latter are increasingly fluent in the vocabulary of biomedicine. Therefore the role of an "interpreter" is progressively less necessary: middlemen without any problem-domain skills or technical competence are eminently replaceable. Mere fascination with technology rarely suffices as a job qualification today: I've known clinicians and researchers with the attention span of 5-year-olds who loved playing with electronic toys, but couldn't do real work with them.

It is shameful that even today, unlike in bioinformatics, programming familiarity is not imparted in many medical informatics programs, and many medical-informatics professionals couldn't write a line of code if their lives depended on it. Informatics professionals who lack basic technical skills—or fail to keep these skills up to date, which ultimately has the same end result—risk coming across like the "Pointy-Haired Boss" (PHB) in Scott Adams' "Dilbert" comic strip. (Incidentally, Adams is a former software developer.)

The PHB's motto: "Anything I don't understand must be easy." In many places, programming is considered intellectual grunt work. I once reviewed a health-informatics paper from a European group describing the use of software that assisted structured clinical note taking. A Google search revealed that the software and its documentation had been created single-handedly by

a lowly (but phenomenally bright) programmer who made it freely available on his (noninstitutional) website. The authors hadn't acknowledged his efforts through coauthorship or even a thank you. In pointing this out, I couldn't restrain myself from being scathing. I shouldn't have needed to aim a gun at the authors' heads to get them to recognize a key intellectual contribution. The programmer had probably left the group, as good people who don't get respect do eventually.

When you learn to program, you realize that programming is a tool that extends the mind. Further, when you try to make a program work, the computer doesn't care who you are: it complains or gives wrong answers if your code is syntactically or logically erroneous respectively. Weinberg [12], writing during the punched-card era, describes people who would turn livid whenever their code was rejected due to their errors. People who can't bear to be contradicted by a machine should leave this business: the emotional strain of repeatedly being wrong will kill them prematurely.

Given the ever-expanding knowledge in the field, you can't be expected to have the technical chops of a fire-breathing programmer, though I've known medically trained folks who have such chops—including my mentors Randy Miller and Dan Masys. Still, nontrivial exposure to programming is necessary if only to instill a sense of humility and respect for what your underlings do. The most quotable line from Clint Eastwood's second "Dirty Harry" movie, *Magnum Force,* is "A man's got to know his limitations." When working with software developers, informatics professionals who don't know theirs may either dictate that their teams use obsolete, inappropriate approaches that are now as dead as the dodo, or worse, demand designs in which the technology to implement them expeditiously doesn't exist. This is like the American tourist stranded in rural Mexico who assumes that, by waving his hands frantically and talking at the top of his voice, he will get the local peasant to understand English. The scene would be hilarious if the clients' money wasn't being wasted because of the failure to communicate.

2.4.1 Minimal technical skillset

The following list constitutes what I believe is the *bare minimum* technical skillset that a professional supporting computing at any level—system designer, project manager, informatician—needs.
- Knowledge of at least one general-purpose programming language—preferably one that allows creation of proof-of-concept prototypes rapidly. Programmer colleagues who may have just started learning the biomedical domain may relate better to a prototype than to an abstract requirement spec. The choice of language has shifted over the years: Perl has been gradually supplanted by Python, and many users of PCs started with Visual Basic. (Its replacement, Visual Basic.NET, is robust enough for creating industrial-strength applications, but somewhat less easy to learn.)
- Knowledge of a variety of microcomputer packages—spreadsheets, databases, word processors, presentation software, drawing/design, project management, website authoring software, and blogging software—at the power-user level, which may include learning the

programming language that is bundled with the package. Visual Basic for Applications (VBA), the programming language of Microsoft Office, is easy enough to be learned by hobbyists.

- Ability to use most features of a microcomputer statistics package and a data-mining package.
- Familiarity with the vocabulary of life sciences and healthcare. (More on this in subsequent sections.)

2.5 GENERAL SKILLS AND BREADTH OF KNOWLEDGE

A quote from George Bernard Shaw's *Maxims for Revolutionists* [13]: "No man can be a pure specialist without being, in the strict sense, an idiot." To be a good researcher, you need to be, at least partially, a generalist. Really good generalists tend to be rare.

> **Maxims, an appendix to Shaw's play *Man and Superman*, was basically Shaw's excuse to collate all the witticisms he had devised over the years but had no opportunity to use in his plays. Parts of it are better-remembered than the play itself.**

General knowledge (in addition to depth in your chosen field) is important when participating in a multidisciplinary field. When you meet with collaborators who approach you for help, you will repeatedly be in unfamiliar domains of knowledge. While you can't know everything about everything, breadth, combined with humility, lets you know what you don't know. Such self-knowledge is the first step to enlightenment—as Francis Bacon said, "He who is content to begin in doubt shall end in certainty."

- The first step is to be honest and admit your ignorance. You will warn your collaborators that your initial questions will be "dumb," but hopefully get smarter over time. You will ask them for references on their area of expertise, starting with introductory works or textbooks, if necessary.
- Even if you realize that it will take far too long for you to master the area enough to deliver a solution to the specific problem, you may be able to find someone—in the geographical vicinity, or elsewhere—with the requisite expertise and persuade them to collaborate. Knowledgeable specialists respect people who have made an effort to learn at least the basics of the specialists' fields and who admit their own limitations openly. Such people are more likely to defer respectfully to the specialist's judgment on matters relating to that specialty. Conversely, someone requesting a specialist's help purely on a business-transaction level is more likely to act as a backseat driver.

Many major breakthroughs have been made by researchers who were aware of advances in other fields and applied them to their own. A famous example is psychologist Raymond Cattell (1905–1998) [14], who applied multivariate statistics, notably factor analysis [15] to the characterization of intelligence and personality. Cattell built upon the work of his friend/collaborator Charles Spearman (creator of the first rank-correlation coefficient). In informatics, Needleman and Wunsch's 1970 sequence-comparison algorithm [16], originally devised to find similarities in amino acid sequences of proteins, has been applied to problems as diverse as 3D reconstructions of images [17] and social behavior of dolphins and bird species as evinced by their vocalizations [18].

Basic knowledge of the biomedical realm is critical to acquire, just as basic knowledge of business or architecture would be necessary for a computing professional supporting those fields. Developers who have worked their way up the ranks have learned that you cannot design a complex system that reflects reality until you've acquired significant knowledge of the problem domain. A talent for acquiring such knowledge quickly, and *enjoying* the acquisition process, also helps. Such talent manifests as a habit of reading widely (especially nonfiction), actively and *critically*–interrelating what you read to what you already know or have read elsewhere. If learning new things doesn't turn you on, you should change careers.

I'm always impressed by trial lawyers who teach themselves a brand new subject at extremely short notice, often well enough to put opposing expert witnesses on the ropes and win their cases convincingly. A story in trial lawyer Louis Nizer's *My Life in Court* [19] describes how Nizer learned about the finer points of music in order to successfully prosecute a copyright infringement case involving a 1945 Andrews Sisters hit tune, "Rum and Coca-Cola" [20]. The good news is that in research, there is usually much less time pressure on you than in law, and the objective is to get closer to the ultimate truth rather than to "win." This is notwithstanding the occasional presence of pathologically competitive researchers: some rivalries, such as the "Bone Wars" [21] between the 19th century dinosaur hunters Othniel Marsh and Edward Cope, became tabloid fodder.

2.6 COMMUNICATION SKILLS

The layperson's stereotype of the programmer as an antisocial recluse and sometimes solitary genius is wildly inaccurate. While geniuses in any field involving sustained mental activity may prefer their own company, building large systems is an intensely *social* activity.
- Programming itself is social: when one writes code, either for the team or as part of an open-source project, the code is meant to be *read* and *understood* by others. If you leave the group, those who have to maintain the system had better find your code comprehensible. Obscurity leads to misunderstandings by others, which leads to defects. Also, one of the standard practices in teams is *code reviews,* during which your colleagues look at your code for possible logical errors. Obscure code is difficult or impossible to review.

   Writing clear code is a form of effective communication similar to the act of writing a scientific paper. Obscurity may have worked for the philosophers Kant and Hegel, and post-modernists may have elevated it to an art form, but it has no place on a team project.

It is possible to write frighteningly obscure code: look at some of the "Coding Horrors" highlighted in McConnell's book [9]. When the language C was introduced on microcomputers in the 1980s, some programmers would purposely write the most obscure C code they could, believing it showed off their virtuosity. On teams, deliberately trying to make others' heads explode when they read your code is frowned upon: James Joyce wannabes are warned once, and then fired.

- Similarly, if you don't communicate clearly with your colleagues and clients, both in speech as well as in writing, the system that your team builds might be a mismatch to the client's needs.
- To get the recognition your team members deserve, they must be able to clearly present the advances they've made to their peers (through papers or presentations). When developers move between jobs, papers and presentations, which are part of the public record, are typically the only things, other than their skills, that they can take with them. Recommendations alone may not count for much unless they are stellar *and* the person making them is well known and highly respected.

  If you insist on being the only one who can speak or write for the group in public forums, you will be correctly perceived as taking credit for your entire team's efforts and obstructing their personal growth. In the long term, as Abe Lincoln said, you can't fool everyone all the time. In the intermediate term, your best people will leave, and your ability to deliver on your commitments will suffer.
- Finally, if you are running a small team with a large workload, the only way to make your load manageable is by training and teaching your users with the hope that most of them will come to you only for major problems rather than minor ones that they could have resolved themselves if only you had taught them a little. Teaching and training are forms of prophylaxis that let you get off the critical path where possible, freeing you to focus on the challenging problems. (If you insist on being indispensable, no matter how small the issue, you're underemployed.)

2.7 MANAGING PEOPLE AND PROJECTS

I assume that you lead the team and are able to either pick most of your team members or have some control in determining which of them to retain. In other words, I'm hoping that your group does not exemplify incompetence or dysfunction at either the individual or team level. The two best books I've read that focus on managing software projects are Steve Maguire's *Debugging the Development Process* [22] and Steve McConnell's *Rapid Development* [23].

Maguire is a former Microsoft manager whose specialty was rescuing troubled projects, and his book is full of war stories as to how they got troubled to begin with. The first sentence in Maguire's book: "This book might make Microsoft sound bad." It is a tribute to Microsoft's self-critical perspective (at that time—things appear to have changed) that they admitted their mistakes and learned from them. (Incidentally, Microsoft Press publishes the book.) McConnell, whose group also provides training to new Microsoft hires, distills his own experience and that of numerous other authors.

It is incredible how much project management is really about managing your people through emotional and social intelligence. DeMarco and Lister state in their classic, *Peopleware* [24] (which is also cited extensively by both the previous books): "The major problems of [software development] are not so much *technological* as *sociological* in nature." In computing, at least in groups with a start-up mentality, most "management" is really support of already motivated people.

"Support" implies shielding your people from higher-level politics and unnecessary busywork (meetings without purpose, writing reports nobody will read), and trusting their judgment and commitment, instead of micromanaging them. It means that, just as you allow an infant to try walking by oneself even though he or she will fall down several times, you must allow your people to fail occasionally, without the fear of being flayed alive. It also means helping them to grow and advance, even if it means that they will ultimately leave your group (the children analogy again). The previously mentioned authors encourage you to think long term: thus, the departure of someone whom you have nurtured, but who leaves on excellent terms with you, means that you have a collaborator and ally, and your group gets the reputation of being one that is a great place to grow so that you attract better talent.

2.7.1 Multifunctional teams

In large teams, there are systems analysts, programmers, testers, a project manager, other leads (lead programmer, lead tester), and an overall lead. Small shops, such as those involved in supporting clinical research, rarely have the luxury of such division of labor. If one person goes on vacation or leaves, that role can come to a screeching halt. Small teams have to be *multifunctional,* with every person potentially capable of filling two or even three roles—even if they routinely perform only one role.

> **I once interviewed an ex-Fortune-500 developer who, while applying for a job, stated that he had no intention of writing code. In one sentence, he had talked himself out of candidacy.**

2.7.2 Software project management

The role of the project manager (PM) is to elicit effort estimates from the team, judge whether the estimates are realistic, keep the project running smoothly by running logistics (supplies, equipment, a separate location and food if necessary, anticipating roadblocks, and preventing them where possible), and providing emotional support. In brief, a good project manager makes it possible for people to work. Ideally, project managers need to have both domain and technical skills: they may be programmers or analysts who moved up. Good PMs are worth their weight in gold.

> **A story from Tom DeMarco in *Peopleware*: As a junior programmer, DeMarco once fell sick, but dragged himself to work in order to give an important product demo. His project manager (Sharon Weinberg, who later became president of a consulting company) saw him, went away, and came back with a cup of hot soup. After eating the soup, DeMarco gratefully asked Weinberg how she found time to do this *and* manage the project. Weinberg smiled and replied: "Tom, this *is* management."**

Some aspects of PM are considered specialized enough that many managers get certified as Project Management Professionals (PMPs). However, PMP certification only indicates book smarts, like a medical or legal degree, not competence, let alone the emotional intelligence that is necessary to do the job well. Also, project management software, which facilitates tracking, is now not significantly more difficult to use than a word processor once you've acquired the basic vocabulary. I've seen excellent PMs who had no special certification, and certified PMs who couldn't manage their way out of a paper bag. Stellman and Greene [25] point out that the PM's most important traits are honesty/transparency, objectivity, and trusting one's team rather than trying to ride herd on them. (Competent programmers tend to be pretty motivated to begin with and do not need the cracking of whips to do their jobs.)

While large teams need PMs, I find it hard to justify the role of full-time PMs on small teams of 10 people or less: the PMs, unless they are also doing systems analysis or programming, will be underemployed, and underemployed folks tend to be trouble, filling their spare time with political games. If you're the lead of a team, you might find it more productive to learn PM software yourself. One word of advice (from Maguire and McConnell): *Don't* use the capability of the software that allows you to specify the percent completion of a task. Most people are notoriously bad judges of percent completion, as in the quote from the late, beloved baseball great Yogi Berra, "90% of baseball is mental. The other half is physical." Sometimes the "last 10 percent" can end up taking 90% of the time. Both authors recommend using *binary* milestones: either a task is done or it's not—and breaking up a larger project into a large number of small milestones, each of which should take about 2–3 weeks to complete.

2.7.3 Effective and ineffective teams: interpersonal factors

A brief passage in Robert Townsend's management book, *Up the Organization* [26], summarizes the 1954 siege of the French garrison at Dien Bien Phu by bedraggled Vietnamese revolutionaries, the Viet Minh. The French ultimately surrendered to a far more poorly equipped force: even bombing raids by 37 US aircraft ordered to their rescue couldn't save them. A French prisoner-of-war later observed that the Vietnamese generals ate the same food as their enlisted men, wore the same clothes, and lived in similar tents, with only a red armband to identify their rank.

The most effective leadership is similarly low key. Lao Tzu says: "As for the best leaders, the people do not notice their existence. The next best, the people honor and praise. The next, the people fear; and the next, the people hate. When the best leader's work is done, the people say, 'We did it ourselves.'"

Multifunctional teams in particular tend to be strongly egalitarian, even more so than software developers in general. Merely wearing a suit and affecting a pompous manner doesn't get you respect—see Dilbert's PHB again—and a truism among competent developers is, "Show me the most formally dressed member of the team, and I'll show you the village idiot." (The only time developers dress up is when they have to meet clients who may expect or enforce a dress code.) Respect has to be earned: for example, some persons on a team with more experience and knowledge than others will be relied on as mentors and coaches and will merit the honorific *Guru* (originally Sanskrit for "teacher, master").

> One of the most effective teams I've personally encountered is at the US National Center for Biomedical Information (NCBI). The Director, Dr. David Lipman, has a challenging task managing a team of very talented people who, he openly states, would easily earn more than double their present salaries if they moved from Government to industry. Welding such talent into a team with a single purpose, and retaining their loyalty, is not easy. If he can't offer significant financial rewards, Lipman compensates by conferring respect and support in numerous ways.

> The weekly meetings (with food provided) are really rallying sessions during which individual members who did anything praiseworthy, no matter how junior they are, stand up and receive accolades. The collective energy is reminiscent of the 2004 Detroit Pistons, the NBA team who convincingly beat the star-studded Los Angeles Lakers 4–1 without a single recognizable "superstar" of their own, through exemplary team play. More important, even when none of his team members are around, Lipman consistently avoids taking credit for individual team members' accomplishments—emphasizing at the *start* of his lectures in a prominent slide, who on his team did the actual work that he is presenting.

As for the worst teams, I've seen a four-tier hierarchy in a team of 10 people, where the suit at the top spent all his time organizing meetings and micromanaging his people. Only the systems analysts were permitted to talk to the client (with resulting delays if they were unavailable or overcommitted). The project manager, a full-time politician lacking both domain and technical skills, was as useful to the team as an inflamed hemorrhoid—a continual irritant to the others.

A programmer title hardly implies social maladjustment or impaired communications skills. A direct programmer-client conversation is often the fastest way to resolve a particular design question. Also, no-direct-communication policies stunt the programmers' growth by limiting their opportunities to learn the problem domain. (You cannot expect them to remain programmers forever.) Finally, if you trust your programmers to write good code, you should also trust their judgment as to which issues are important enough to keep you in the loop.

For numerous examples of both intrateam harmony and dysfunction, read *Peopleware*.

2.7.4 Exploitative practices: a selfish reason not to employ them

Research, especially in the academic world, is often characterized by low pay, with extensive use of students as part-time staff. While apprenticeship of students may be justified—as long as you are genuinely taking the effort to mentor them, as opposed to just using them as an extra pair of hands—I've also seen places where this practice extended to exploit fresh graduates or average people looking to earn a living.

Unless the work is completely mechanical and noncerebral, such policies are short sighted. If you pay your research assistants like burger flippers, don't be surprised if the average duration of employment is 6 months. One Yale researcher I knew paid unwed welfare mothers minimum wage to perform secondary data entry from paper documents from home. While claiming to perform a social service, he didn't have a charitable bone in his body. He was only being cheap, avoiding paying market wages to data-entry professionals.

The problem is that when new staff is hired, they can't function productively until a couple of months have passed, and the time of (relatively expensive) senior personnel is consumed with training and babysitting them. This chore takes the senior staff's time away from their own work. If this happens over and over again due to staff departures, the project gets back-logged: it would have cost much less to pay the serfs a decent wage to begin with. In the case of the mothers—well, let's just say that you can't expect better than atrocious data quality with secondary data entry—already a dubious practice, as I'll discuss in a future chapter—from distracted, sleep-deprived, multitasking women who are concurrently managing needy (by definition) infants.

Exploitative practices always have unintended consequences in almost all academic fields. Adjunct faculty is often similarly abused, with PhDs earning $18,000/year even after teaching a full course load, and analyses show that university presidents' salaries correlate with the proportion of staff hired as adjuncts [27,28]. This approach backfires after the good adjuncts seek jobs elsewhere, while the indifferent ones realize that the easiest path to continued employment (which is based mainly on student evaluations) is easy and inflated grading of students rather than conscientious imparting of knowledge. (You can get away with studying as little as possible, if I can go through the motions of pretending to teach.) Years later, after applying for jobs and demonstrating their ignorance of subject matter to their prospective employers, the students realize too late that they were short changed in the Faustian bargain.

2.8 PERSONALITY TRAITS

The penultimate chapter of Steve McConnell's *Code Complete* [9] follows the lead of the noted computer scientist Edgser Dijkstra, whose 1972 Turing Award speech was entitled "The Humble Programmer" [29]. Both the book and the speech should be on your essential-reading list. Dijkstra emphasizes that both computers and programs are now so large and complex that the only way we can make them manageable is by being aware of the limitations of the human mind.

Some of the traits that make a great developer may be inferred from this chapter's earlier text, but I'll summarize the essential traits here: intelligence coupled with humility, curiosity and the desire to learn, intellectual honesty, discipline, creative laziness, good communication, and a cooperative attitude. With respect to the last trait, while individuality is celebrated—because each team member brings a novel perspective—*prima donnas* are rarely tolerated on teams: whatever talents they may have are more than offset by the team's disruption. Interestingly, *persistence* is overrated: good developers know when to stop banging their heads against walls, take a break, and then change strategies.

One trait I might add to McConnell's list, with respect to clinical research, is *minimalism*. That is, given your team's limited resources, most of the solutions that you devise need to be *good enough*. Any additional effort invested in "ld-plating"—adding a whole bunch of features the client didn't request—rarely pays off.

2.8.1 Personality profiling: a word of warning

At one time, psychological screening to identify the "programmer personality" was a focus of investigation. While *aptitude* tests may have merit, psychological profiling is dubious. For example, programming talent implies nothing about morality: there are highly ethical software developers as well as career criminals who break into systems for profit.

> Research using the Myers–Briggs Personality Inventory (inspired by Carl Jung's work) showed a predisposition of the Introversion, Thinking, and Judging (ITJ) personality subtype, where "Introversion" means interested more in the inner world of ideas than in the external world of things. Myers–Briggs, however, is now known to have poor intertest reliability (a person's "type" may change over weeks or months), and is regarded by many, for example, management professor Adam Grant [30], as pseudoscience similar to Freudian psychoanalysis. It is worthless when applied to *predict* job performance.
>
> In any case, you don't have to *enjoy* schmoozing to be an effective communicator, any more than an expert martial artist capable of crippling or killing an adversary necessarily enjoys violence. A psychiatrist (and martial artist) friend of mine administered a questionnaire measuring aggression levels to several of his fellow martial artists. The data showed a statistically significant *inverse* correlation between aggression levels and rank (belt level) attained. My friend's *sensei* (teacher), a tough-as-nails 5th Dan revered by his students for his nurturing personality, had by far the lowest level. Of course, correlation doesn't prove causation: the *Sensei* may have weeded out the pathologically violent types so they didn't make it to the higher levels.

2.9 NEGOTIATION SKILLS

In an academic center, the fact that your team exists to support research doesn't mean that they can't be compensated fairly for their efforts. While your team may not break a "profit," you at least have to break even. Most federal agencies now insist that support groups have cost-recovery mechanisms in place. Five people can't do the work of 20, and if you're not compensated adequately, you can't hire the extra people you might need, nor can you retain your soon-to-be-overworked people. The following passage is a guide to negotiation with people who may become your collaborators for the next 3–5 years. As in the case of a marriage, you need to tread cautiously.

I wish I had a buck for every time I've been approached by someone *after* their proposal got funded federally, who then proceeded to tell me what they needed, and then stated that they had a limited budget for support. In most cases, this is the equivalent of walking into a fancy French restaurant demanding the nine-course meal with a different wine after every course, while stating that you can only afford to pay for a burger and fries. The late Professor Kathryn Chaloner of the University of Iowa Department of Biostatistics gave me some sage advice on the carrot-versus-stick approach to preventing such situations, while at the same time avoiding having to tell such folks plainly to go to hell.

You set an hourly rate for the services of various members of your team that is *at least* double their hourly wage-plus-benefits (WPB) rate. You announce your group's policy that anyone who contracts with you *after* they were funded will pay full price, and that your team will only do as much as they can before the budget runs out, with no assurances that the work will be completed. This is as effective a disincentive as any with respect to others' taking your team's support for granted.

However, anyone who does you the courtesy of contacting you *in advance* (when the proposal is being planned) will get a rate equal to the WPB rate. If the commitment of any team member is extended, it may make more sense to budget for a portion of their salary. This approach allows you to project your group's incoming revenue and resource needs over the next several years, and budget accordingly—thus, you may need to hire extra people if demand exceeds supply. More important, you may be able to do pilot work that will strengthen the proposal and make it more viable. (The added incentive is that your team commits to such pilot work being done gratis: after all, if you share in the rewards of a funded proposal with the investigator, your group must also share part of the risk of having your efforts come to nothing.)

2.10 CHOOSING YOUR COLLABORATORS

However, choosing who you work with is not a simple matter of dollars and cents. People in academia are neither more nor less considerate of their fellow human beings than those outside it. I've known some who are both sterling scientists and human beings, and it is both a pleasure and honor to collaborate with such people and learn from them. I've also known narcissists and sociopaths who exploited their power and their underlings. Even in a strict cash-for-service transaction, such people will be trouble: they'll find ways to demand more service without paying for it, will insist on telling you and your people how to do their jobs, and treat your team members abusively. If you have the freedom to decline a partnership with such people, use it. If you can't say "no" directly, employ an excuse such as overcommitment, or quote an obscene price for your team's participation: this will send them elsewhere.

The first step to undertake when somebody approaches you for support is to never say yes or no after the first meeting, but promise to get back to them soon. You then research them, preferably by asking trusted colleagues, as well as via Internet. One way is to research their past graduate students or postdocs, and try to find out whether these folks, who have now moved elsewhere and possibly have faculty positions themselves, are still collaborating with them, as evinced by joint papers published well after that person moved on. If their lives as underlings were a living hell, you can be sure they won't be collaborating. (By contrast, one of my mentors, Professor Charles Bruce, a Yale neuroscientist, has ex-students who are all over the globe and who, like me, love him; he writes his own software for brain-signal recording and sends updates to them.) You may also find out, from the Institution's records, how long, on average, their students took to graduate. One student taking an unreasonably long time may suggest that the problem is with the student. If almost *all* students take long to graduate, however, it suggests that their adviser is either neglecting them, exploiting them, or both.

2.10.1 Supporting junior faculty

Junior faculty members just starting their careers do not have access to much money, so you and your team have to switch, on occasion, from cash-for-service mode to *pro bono* work. Without your participation in preliminary work, they may not have material with which to write papers or persuasive grant proposals. Your budget should make sufficient allowance for such work, which should be regarded as an investment in the future—if a successful grant proposal results, your team participates in the bounty.

Still, there are some safeguards that need to be in place before you undertake such commitments.

• Make sure that the researcher is sufficiently committed in terms of putting in at least as much work into the project as your team will be doing—for example, with literature gathering and a systematic plan of attack for the problem. While your team may be doing the work for free, they must be paid back in kind, through coauthorship or acknowledgment in the eventual publication of the work.

   If researchers haven't invested enough energy to have an emotional stake in the project's success, the work will never see the light of publication. More important, researchers who haven't expended energy can simply change their minds and walk away after having wasted everybody else's time.

   One way is to know whether such commitment exists is to ask the client for a brief write-up, which includes a citation of relevant literature and how the present work plans go beyond what is already known. Explain that you need this background to educate yourself and help the researcher more effectively. It is interesting how many people who approach you with a verbal request for help will never show their faces to you again.

   Institutional Review Boards (IRBs) typically insist on similar due diligence anyway (as, obviously, do grant review committees). Don't rely on the IRB, however, to filter half-assed science for you. Their role is to protect human subjects. If the proposed work involves a retrospective data analysis—for which your team does the considerable data gathering/massaging—the IRB is not necessarily the best judge of a clueless research plan or work that will go nowhere.

I've known junior faculty who, after getting commitment from statistics or computing collaborators, will then delegate everything to lowly research assistants. One researcher didn't even show up for meetings that were organized at *her* request, repeatedly dropping out at the last minute. Department chairs may have legitimate excuses for this: assistant professors starting their careers don't because they are the only ones who can satisfactorily answer any questions that come up. Research assistants, who may lack the necessary domain knowledge, can't be expected to speak for them. Here, my statistician friend and I decided that her time wasn't more valuable than ours: we put the work on ice.

- Junior researchers may be as inconsiderate or abusive as established ones—though, as Lincoln observed, unpleasant character traits may not be fully apparent until someone is given a little power. Make sure that a written agreement exists regarding which of your team members will be given credit, and how.

I've seen a case in which, after 8 months of *pro bono* work, neither the team member nor the member's Departmental Division was acknowledged in the resulting paper, which the team member found out about after the authors' press release.

2.10.2 Scoping the project: change control

Once you've accepted someone as collaborator, it is important that you are able to meet your commitments. However, you need to explain to this person that the process of software development works similar to building a house: if the builder agrees to build a three-bedroom Ranch, then after the contract is signed, the client can't turn around and ask for the Taj Mahal.

Certainly, modest requests may be accommodated, as may requests that are vital due to last-minute legislative/business-climate changes over which the client had no control. However, if the suggestions for changes keep piling up, your effort estimates must change accordingly—and the client needs to foot the bill for the increased development costs. A clear memorandum of understanding regarding such issues needs to exist.

One way to mitigate risks is to have the client pay a modest amount for a prototyping phase that helps the requirements be fully fleshed out. At the end of this phase, a more reliable estimate of actual costs is made. The memorandum should state clearly that, after the prototyping phase is over, any changes that are suggested will not be entertained unless the factors that triggered the change request were beyond the client's control *and* the changes are critical to functionality.

McConnell's books, cited earlier, provide a good primer on change control issues: Chapter 14 of [23] particularly warns against "feature creep." Certain well-known IT disasters in the Federal Government, where projects were abandoned after running way over budget and timeline, were traceable to the complete absence of focus by the clients, who, on the few occasions that they had made up their minds, kept changing them with the phases of the moon. Candice DeLong's insider account [31] of the issues surrounding the failure of the computerization overhaul at the Federal Bureau of

Investigation—which included a 4 h meeting to decide whether to use green or blue as the background color for the program—is comical until you remember that taxpayer dollars were wasted because of the dithering.

> **Habitual indecisiveness can have lethal outcomes: McConnell cites the 1991 movie *Bugsy*. Mobster Bugsy Siegel—responsible for building Las Vegas's first casino, which cost six times more than originally estimated due to feature creep—was executed by his Mafia colleagues who ran out of patience after Bugsy changed his mind once too often.**

Collaborators with attention deficit disorder—and these are as numerous in academia or industry as in government and the Mob—spell trouble. They will consume a disproportionate amount of your team's resources, not to mention driving them up the wall and considering job opportunities elsewhere. Even an open-ended hourly billing model for service may not help: such clients will find a way to blame and bad-mouth you for the delivery delays that result from problems of their own making, and life is short enough without spending time doing damage control with regard to your other clients.

If you see signs of ADD during the prototyping phase—and this can manifest in little ways, such as the collaborator taking a disproportionate amount of time to respond to simple queries or toggling their feature requests (asking for a feature, then, after it has been added, asking that it be removed for no good reason), you would do best to cut your losses and end the collaboration as soon as possible, using whichever excuses you can devise. If you have to write off your team's unreimbursed prototyping fee, so be it: to use a marriage analogy, the sunk cost of an engagement ring is a small fraction of a divorce settlement.

2.11 TOPICS IN CLINICAL RESEARCH SUPPORT

2.11.1 Special aspects of supporting investigational drug studies

Your institution may be involved in FDA-regulated studies for investigational therapies or devices, sometimes in partnership with a commercial organization. If the study is done as part of a partnership, the level of informatics support that you may have to provide may be, paradoxically, much *less* than with investigator-initiated or federally funded studies.

Some of the issues can be traced to the FDA regulations regarding the software that can be used for such studies. The Code of Federal Regulations (CFR) document that dictates the standards with which such software (*and* the organization employing it) should comply is known as "Title 21 CFR Part 11" [32]. The essential aspects of CFR 11 deal with the following [33]:

- *Software aspects*: The system must authenticate each user (through login). Every record must have an electronic signature of the person/s who created or modified it. The software must maintain a detailed audit trail of the creation, modification, and deletion of every record (which user did what and when).

- *Organizational aspects:* The organization must properly document the individuals responsible for creating the data in the system used, and must document all standard operating procedures (SOPs) for all operations related to CFR 11, with periodic process reviews to assess adherence.

With the advances in technology, the software aspects are not too challenging to implement: there are numerous, affordable third-party toolkits that support implementation of each aspect individually. A more important stumbling block relates to validation of the software, which must be done by a third party, whose services are not cheap (read, multiples of $100,000). For open-source systems such as REDCap [1], which are essentially given away freely, there is little motivation for the creators to invest this kind of money when they have no hope of recuperating it.

> **The FDA has not exactly had a reputation for clarity. A good overview is provided in the Wikipedia article [34]; the account of the history of CFR 11 is disheartening. For example, some of the revised 2003 regulations contradicted those of the "Final" 1997 rule. The FDA-issued "Guidance for Industry Computerized Systems used in Clinical Investigations" in May 2007 [35] overruled the same document of 1999. The planned revision of CFR 11, originally scheduled for 2006, has been pushed back into the indefinite future.**

> **Overly complex or prescriptive legislation—the US Federal Tax Code, Sarbanes–Oxley, the Americans with Disabilities Act (ADA), and the Affordable Healthcare Act—even if originally well-intentioned—ends up creating an industry of lawyers to deal with their nuances. See *The Economist*'s take on the 848-page Dodd–Frank legislation [36]. The ADA has also spawned a plague of predatory "drive-by" lawsuits targeting small-business owners for minor violations [37].**

> **A boon to commercial software also results. TurboTax™ alone saves taxpayers millions of dollars in aspirins and antihypertensive medication. Several commercial software packages exist only to help an organization deal with all of the FDA's paperwork and documentation requirements. Open-source software creators find the task of creating software to deal with byzantine (and often continually mutating) legal issues uninteresting both technically and scientifically.**

As I stated in the previous chapter, the fact that particular software is "validated" generally only implies that the software meets its specification, and that its creators have followed the letter of the law. It implies nothing about its usability or functionality, because the specification itself may be flawed. Also, the regulations state almost nothing about security, which I discuss in chapter: Computer Security, Data Protection, and Privacy Issues.

Many of the CFR-11-compliant commercial systems that support data capture for regulated studies do not interoperate well with other systems: simple bulk import and export of data may be a complex programming task. Therefore, when academic institutions collaborate with organizations that use these, a disproportionate amount of the workflow is paper based, with data being filled in on forms supplied by the sponsor. Also, the commercial systems are pricey enough (think $10 M plus) that, unless an institution is performing a large number of regulated studies—for example, as part of an active drug or device development program, with a view to commercializing the resultant intellectual property—an investment in such a system is rarely cost effective. If your institution has purchased such a system, of course, babysitting it will be part of your team's responsibilities.

BIBLIOGRAPHY

[1] P.A. Harris, R. Taylor, R. Thielke, J. Payne, N. Gonzalez, J.G. Conde, Research electronic data capture (REDCap)—a metadata-driven methodology and workflow process for providing translational research informatics support, J. Biomed. Inform. 42 (2) (2009) 377–381.
[2] I. Somerville, Software Engineering, 10th ed., Pearson, London, UK, (2015).
[3] F. Brooks, The Mythical Man-Month, and Other Essays, 20th Anniversary ed., Addison-Wesley, Reading, MA, (1995).
[4] K. Wiegers, Software Requirements, Microsoft Press, Redmond, WA, (2003).
[5] S. McConnell, Software Project Survival Guide (Developer Best Practices), Microsoft Press, Redmond, WA, (1997).
[6] A. Stellman, J. Greene, Learning Agile: Understanding Scrum, XP, Lean, and Kanban, O'Reilly Media, Sebastopol, CA, (2005).
[7] S. Yegge, Good Agile, Bad Agile, Steve Yegge, Kirkland, WA, (2006).
[8] Wikipedia. Agile software development. Available from: https://en.wikipedia.org/wiki/Agile_software_development, 2015.
[9] S. McConnell, Code Complete, 2nd ed., Microsoft Press, Redmond, WA, (2004).
[10] J.W. Cooley, J.W. Tukey, An algorithm for the machine calculation of complex Fourier series, Math. Comput. 19 (90) (1965) 297–301.
[11] S. Alinsky, Rules for Radicals: A Pragmatic Primer for Realistic Radicals, Random House, New York, NY, (1971).
[12] G.M. Weinberg, The Psychology of Computer Programming, Silver Anniversary ed., Dorset House, New York, NY, (1998).
[13] G.B. Shaw, Maxims for revolutionists. in: Man and Superman. Available from: Bartleby.com, (Ed.), 1903.
[14] Wikipedia. Raymond Cattell. Available from: https://en.wikipedia.org/wiki/Raymond_Cattell, 2015.
[15] D.J. Bartholomew, F. Steele, J. Galbraith, I. Moustaki, Analysis of Multivariate Social Science Data, Statistics in the Social and Behavioral Sciences series, 2nd ed., Taylor and Francis, London, UK, 2008.
[16] S.B. Needleman, C.D. Wunsch, A general method applicable to the search for similarities in the amino acid sequence of two proteins, J. Mol. Biol. 48 (3) (1970) 448–453.
[17] S. Knowles-Barley, N.J. Butcher, I.A. Meinertzhagen, J.D. Armstrong, Biologically inspired EM image alignment and neural reconstruction, Bioinformatics 27 (16) (2011) 2216–2223.
[18] A. Kershenbaum, L.S. Sayigh, V.M. Janik, The encoding of individual identity in dolphin signature whistles: how much information is needed?, PLoS One 8 (10) (2013) e77671.
[19] L. Nizer, My Life in Court, Doubleday, New York, NY, (1961).
[20] Wikipedia. Rum and Coca-Cola. Available from: https://en.wikipedia.org/wiki/Rum_and_Coca-Cola, 2015.
[21] Wikipedia. Bone wars. Available from: https://en.wikipedia.org/wiki/Bone_Wars, 2015.
[22] S. Maguire, Debugging the Development Process, Microsoft Press, Redmond, WA, (1994).
[23] S. McConnell, Rapid Development, Microsoft Press, Redmond, WA, (1996).

[24] T. DeMarco, T. Lister, Peopleware: Productive Projects and Teams, Dorset House Publishing, New York, NY, (1997).

[25] A. Stellman, J. Greene, Applied Software Project Management, O'Reilly Media, Sebastopol, CA, (2005).

[26] R. Townsend, Up the Organization: How to Stop the Corporation from Stifling People and Strangling Profits, 35th-Year Commemorative ed., Alfred Knopf, New York, NY, (1970).

[27] L. McKenna. The College President-to-Adjunct Pay Ratio. The Atlantic 2015 9/24/2015. Available from: http://www.theatlantic.com/education/archive/2015/09/income-inequality-in-higher-education-the-college-president-to-adjunct-pay-ratio/407029/, 2015.

[28] M. Wood, A. Erwin. The One Percent at State U: How Public University Presidents Profit from Rising Student Debt and Low-Wage Faculty Labor. Available from: http://www.ips-dc.org/one_percent_universities/, 2014.

[29] E.W. Dijkstra, The humble programmer (Turing Award lecture), Commun. ACM 15 (10) (1972) 859–866.

[30] A. Grant. Say Goodbye to MBTI (Myers–Briggs Type Indicator), the Fad that Won't Die. Available from: https://www.linkedin.com/pulse/20130917155206-69244073-say-goodbye-to-mbti-the-fad-that-won-t-die, 2013.

[31] C. DeLong, Special Agent: My Life on the Front Lines as a Woman in the FBI, Hyperion, Los Angeles, CA, (2001).

[32] FDA's 21 CFR Part 11 Electronic Signatures & Records. Available from: www.21cfrpart11.com, 2002.

[33] C. Ruggles. Is your Electronic Lab Notebook Software 21 CFR Part 11 Compliant? Available from: http://www.kinematik.com/blog/is-your-eln-software-21-cfr-part-11-compliant, 2014.

[34] Wikipedia. Title 21 CFR part 11. Available from: https://en.wikipedia.org/wiki/Title_21_CFR_Part_11, 2015.

[35] Federal Drug Administration. Guidance for industry computerized systems used in clinical investigations. Available from: http://www.fda.gov/OHRMS/DOCKETS/98fr/04d-0440-gdl0002.pdf, 2007.

[36] The Dodd–Frank Act: Too Big not to Fail. The Economist 22012 2/18/2012. Available from: http://www.economist.com/node/21547784, 2012.

[37] L. Mumma. Lawmakers band together to fight drive-by ADA lawsuits. Available from: http://www.kcra.com/news/lawmakers-band-together-to-fight-driveby-ada-lawsuits/31511266, 2015.

CHAPTER 3

Core Informatics Technologies: Data Storage

3.1 TYPES OF DATA ELEMENTS: DATABASES 101

Individual readers may find some material in this and other sections basic. I've provided sufficient landmarks in the text (ie, subheadings, emphasized words) to allow skimming and skipping as necessary.

Data elements can be categorized into the following types.

1. *Simple data types:* short text (such as Patient Name), integers (eg, number of children), decimal numbers (eg, annual income), date/times (such as date of birth, or appointment date/time). Integers are also used to represent categorical data such as race: here, each integer corresponds to a specific value that is associated with a text description (eg, 1 = Caucasian, 2 = African-American, etc.). Data comprising entirely of simple data types are called *structured*.
2. *Complex (or unstructured) data* include arbitrary-sized unformatted text (eg, an abstract of a paper, or the paper itself) or signal data (eg, EKG data, or images, which are nothing but two- or three-dimensional signals, with time as a dimension for some data). The fields of information retrieval (IR) and natural language processing (NLP) deal with text.
3. *Semistructured elements* are textual but have an internal structure where units of text are surrounded by arbitrarily hierarchical markup. For example, in hypertext markup language (HTML), text in web pages is enclosed between symbols (*tags*) such as <B>...</B>, <I>...</I> to indicate bold/italic formatting or <TITLE>...</TITLE> to indicate content of a Title section. (The "</" symbol indicates "end of.") Extended markup language (XML) [1] applies the same principle to impart a kind of structure to data, as in Fig. 3.1a. JavaScript Object Notation (JSON), so called because it is the way that JavaScript, the programming language of web browsers, represents data internally [2] uses curly braces and prefixes with colons instead, as in Fig. 3.1b, which represents the same data as Fig. 3.1a.

Both XML and JSON are used to represent arbitrarily hierarchical data. XML can, in fact, be used to represent HTML. However, the tags in XML are case sensitive, while the tags in HTML are not.

> **Microsoft Office now uses application-specific XML for all of the Office applications. The .docx format of Microsoft Word, for example, builds on XML files that have been compressed using the popular PKZip algorithm. Try making a copy of a Word document, renaming the copy's extension to .zip, and then double click on it. You will see a set of folders, all of which contain XML files. These can be inspected in Notepad or any other text editor that you have set up as your default XML editor. You can experiment similarly with Excel or PowerPoint files.**

Clinical Research Computing. http://dx.doi.org/10.1016/B978-0-12-803130-8.00003-8

```
<Patient>                                           {"Patient"
      <Name>John Smith></Name>                      { "Name": "John Smith",
      <Birth_Date>1988-01-02</Birth_Date>                "Birth_Date": "1988-01-02",
      <Gender> Male </Gender>                            "Gender": "Male",
      <Address>                                          "Address":
            <Street> 140 Elm St. </Street>               { "Street": "140 Elm St.",
            <City> Peoria </City>                          "City": "Peoria",
            <State>IL </State>                             "State": "IL"
      </Address>                                         }
</Patient>                                          }
                                                    }
(a)                                                 (b)
```

Figure 3.1 *A representation of patient demographic information in (a) XML and (b) JSON.*

All three types of elements can coexist. Thus, to describe a radiology examination, we have patient ID, date/time of examination, hospital location (all structured), the radiograph itself (an image), and a textual description of the findings (either unstructured or semistructured, depending on the software used).

3.1.1 Organization of data: relational databases

A *database* is an organized electronic collection of data. A *database management system* (DBMS) is software that helps you create and maintain databases. Today's commonest DBMS technology, termed *relational database management systems (RDBMSs),* was conceived by the computer scientist Edgar F. Codd, then at IBM, in a seminal, if hard-to-follow, 1970 paper [3].

Codd's penchant for mathematical notation and writing in the abstract induced brain cramp in the average reader and delayed the wider acceptance of his idea. His friend and colleague Chris Date was largely responsible for popularizing Codd's ideas by translating them into an easily understandable form through the extensive use of examples. His textbook on relational database design, now in its 8th edition, continues to be a valuable reference for the advanced practitioner, though it is possibly not the book that you should start with if you know nothing at all about databases.

The relational model organizes all data (conceptually) into tables. A table contains *rows* (or *records*) and *columns* (or *fields*), where each field for a given row is a data element. In production relational databases, there may be hundreds to thousands of tables. Individual tables are logically linked to other tables. The logical links are called *relationships*. Ideally, relational databases allow a specific data item to be *only stored once*, and then *looked up* as needed. Because of this, relational databases make it easier to store data with minimal redundancy.

To be considered relational today, a DBMS must allow databases to be created and manipulated using a language called *Structured Query Language* (SQL) [4]. Originally developed at IBM, the standard definition of SQL is now maintained by a committee within the International Standards Organization (ISO). The advantage of the relational data model is its conceptual simplicity. Microcomputer end-user-oriented databases, such as Microsoft Access, also utilize the relational approach. End-users with little or no programming experience can construct databases, build form- and report-based interfaces to them, and query their data using graphical tools: the query tools generate SQL behind the scenes. (Certain complex queries, however, lack graphical counterparts and must be typed in.) This ease of use, combined with power, is a major reason for the technology's dominance.

3.1.2 Key columns

A table has at least one structured element. (Historically, all elements in a table were structured: support for unstructured and semistructured elements came later.) A structured field, or less commonly, a *set* of structured fields, whose values uniquely identify a given record is called a *Key*.

Most commonly, to designate a key, one employs a special type of column, called an *auto-number* or *identity* column, which contains a sequentially increasing integer generated by the software. In many EHRs, "patient ID" fields are typically based on identity columns. The values of identity columns are not editable by users, and once set by the software for a given row, they can never be changed. Further, they are not recycled when a record is deleted, so that there may be gaps in the sequence; gaps, however, are perfectly acceptable, since the IDs have no intrinsic meaning.

3.1.3 Why employ auto-number columns for keys?

Even though more than one set of fields within a table may suggest itself as a key, only one of these must be used for this purpose: this set is called the *primary key*. When replicated in other tables, the replica is called a *foreign key*. *Primary-key/foreign-key pairs are the basis for logical relationships between tables.* I provide an example next in the context of the electronic health record.

In a Patient Demographics table, we have a patient ID (the primary key), first name, last name, date of birth, social security number, address details, etc. In a Visits table, we record the Patient ID to identify the patient to whom a given visit applies. Therefore, in the Visits table, the Patient ID serves as a foreign key. The other patient-associated demographics fields, such as first and last name, are *not* replicated in the Visits table: instead, you *look them up* in the Demographics table when needed, using the Patient ID to search for the record where they are stored. The mechanism of lookup typically uses indexes, described in the Section 3.3, and can be very fast.

> Note that even though a patient's social security number is unique, we normally prefer not to use it as the primary key, for at least two reasons.
>
> • We cannot determine the social security number of an unconscious "John Doe" who is brought into the hospital when we need to create a record for him in the EHR.

- **The social security number is considered personal health information (PHI). Using it for a primary key implies that multiple other tables—such as a visits table, or a laboratory tests table—would also have to store it in order to refer to specific patients. It is best to localize PHI to a single table, where it can undergo an extra layer of protection, such as encryption.**

Even though the word "key" is singular, the number of columns that constitute a table's primary key can be more than one. For example, in a Visits table, the combination of Patient ID + date/time of the visit *might* serve as a primary key, because a given patient cannot have two visits with the same date/time. However, one prefers to use an auto-number Visit ID as the primary key instead. The reason for this is that (as I now discuss) the more compact the key, the less space needed to store it in multiple places, and the faster the search for a given key value. In other words, *if a primary key's value is needed elsewhere, try to use a single field as a primary key, creating one artificially if necessary*.

3.1.4 Database schemas

The word *schema* is used for the detailed description of a database, which is stored within the database itself in a computable form (ie, it can be read and manipulated by software). Such a description includes definition of individual tables, the names and data types of each column, and the relationships between tables. The schema also stores *constraints* on the values within individual columns, such as primary keys, uniqueness, and maximum/minimum permissible values. Some constraints apply to relationships: for example, a Patient Demographics record cannot be accidentally deleted if Visits for that patient exist in the database.

In relational databases, the schema itself is stored in special tables (called *system tables*) that cannot be manually edited, and which are normally invisible to the end-user—though program code may read the contents. The contents are changed through data-definition statements in SQL, or when the user defines or alters tables, columns, and relationships using the software's graphical user interface.

Database schemas have some overwhelming advantages (which apply most of the time) and some modest disadvantages (in relatively uncommon circumstances). A good overview of the issues is provided by Martin Fowler.

3.1.4.1 Benefits of a schema
1. It can be rendered graphically as a diagram, as well as a textual description. Therefore, *the schema constitutes basic documentation*. Inspecting a schema is the first step toward understanding a database created by someone else.
2. The schema is *active*: that is, it constantly operates behind the scenes, being consulted by the DBMS software continually to ensure data consistency. By specifying constraints in the schema, the programming burden of the database developer is reduced dramatically. The constraints are rules that specify conditions that must hold for the data to be consistent. Constraints are highly concise. Expressed in a few words, they do the work of pages of programming code. The

constraints are *centralized* (ie, they are accessible to all applications that access the database). Also, unlike code, constraints cannot be subverted or bypassed (accidentally or maliciously) by users or by program code. Overriding a constraint can only be done (by someone with administrative-level access) by deleting (dropping) it.

3.1.4.2 Drawbacks

The major drawbacks of schema-based approaches for defining data elements relate to inflexibility.

1. Very often the system's requirements evolve rapidly. (An example of this is software that needs to deal with continually evolving regulations, such as the tax code. The individual rules related to the tax code often rely on data elements that change.) These requirements must be accommodated by redesigning the schema (eg, adding or modifying table definitions). When a table needs to be modified, applications using that table must be suspended or terminated, the modification completed, and then the applications must be restarted. For databases with a vast number of concurrent users, this means, in practice, that the entire database must be taken offline for any modification to take place. Frequent downtime and service interruptions are usually unacceptable.

2. Sometimes, end-users must be allowed to define their own brand new custom columns. At the same time, it is risky to allow them unfettered access to the entire schema because it is impossible to guarantee that a naïve user might not accidentally make a change in the existing tables that causes the application to break. In EHRs and systems that capture clinical research data, brand new experimental parameters (or information on an emerging disease) that were recently discovered (and which the software vendor cannot possibly be expected to be aware of) may need to be captured.

3. In certain situations, the types and number of columns that are necessary to record detailed information are highly variable. Such variation is known for products, where the details depend on the product category. The special-purpose columns necessary to fully describe a product differ depending on whether the product is a food item, a book, a household appliance, computer software, or a vehicle. One can end up with a vast number of tables, one for each category, or a single table with a vast number of columns, most of which are inapplicable for a particular category. Maintaining these tables can become a major challenge.

3.1.5 "Schema-less" design approaches

There are two ways to allow redesign flexibility.

3.1.5.1 Entity–attribute–value modeling

(I've written extensively on this subject in the past: my previous book [5] has a large section on this topic. Skip this section if you are familiar with the subject.)

This is an approach that allows flexibility while at the same time preserving the benefits of schemas (ie, it is used selectively for only a part of the overall system). A description of the approach is provided in Dinu and Nadkarni [6]. Here, we still employ an RDBMS, but use a table with (conceptually) three columns.

- The *Entity* is the "thing" that we are describing. In the context of clinical medicine, it refers to the person whose data is being recorded and the date/time the data was collected.

- The *Attribute* is an aspect of the thing described. In medicine, it refers to a clinical Parameter (eg, a lab test or history/clinical finding). The Entity and Attribute together form the Key of this table.
- The *Value* refers to the value of that parameter (eg, hemoglobin of 14.0 gm/dL, body temperature of 39.8°C).

Since Value columns in most relational database must have a fixed data type, we typically employ multiple EAV tables, one each for values of type text, date, integer, image, etc. (Some systems employ a single table with separate text, numeric, or date fields, only one of which is populated for a given attribute.)

The EAV approach is useful when the data is *sparse*. In clinical medicine, of the thousands of parameters that are defined across all specialties, only a relatively modest number apply to an individual patient. By analogy with the business world, of the thousands of brands of items in a supermarket, most people buy only a couple dozen items, and so a supermarket receipt only records what is purchased. (The attribute is the product, while the values are the quantity, unit price, and total amount paid). There is no purpose in creating columns for the thousands of products that a given customer might have purchased, but did not—or every ailment a given patient could have suffered from, but didn't.

Similarly, when products are introduced or withdrawn, creating/deleting columns for each product would cause unacceptable downtime. It is simpler to edit entries in a Product Descriptions table. The same reasoning applies to novel or obsolete clinical parameters.

The EAV approach is also useful in geographical information systems (GISs), discussed briefly in Section 3.3.2.5, where the attributes associated with a location are highly variable. A location could represent a store, a hospital, a school, or a restaurant, with attributes specific to the category (eg, hours open, cuisine, type of goods sold, public or private school, etc.).

In principle, *all* data could be modeled as EAV. However, it does not make sense to use it for *nonsparse* data such as Patient Demographics, where *every* patient has a name, date of birth, gender, ethnicity, home address, and so on. Creating attribute–value pairs for a given patient such as "Gender – female" and "Date of Birth – 1/1/1988" would take up extra space for the attribute for every patient in the system and complicate the process of data retrieval, with no discernible benefits. Instead, the "*traditional*" approach of using explicitly named "Date of Birth" and "Gender" columns is far simpler and is to be preferred—which is the reason why EAV modeling is not typically taught in introductory database courses.

In databases that use EAV approaches, EAV tables that hold sparse data *coexist* with traditionally modeled tables that hold dense data. The traditional tables typically outnumber the EAV tables by at least 10:1. Most important, the EAV tables must be supported by an Attribute Descriptions table that stores detailed definitions of every attribute in the system—its name, description, data type, constraints, and so on. Programming code consults this table, and prevents errors such as assigning text data to a numeric field, or entering out-of-range values. (In the business scenario mentioned earlier, the Products table is an example of an Attribute Description table.)

The attributes may be grouped into higher-order categories, such as "Forms" that present specific attributes in a particular order. In other words, "Forms" and "Attributes" *simulate the tables and columns of conventional database design approaches.* However, since form and attribute definitions and constraints are created as *data*, the software application does not need to be halted and restarted when these are modified.

The EAV approach is not intended for database neophytes. It requires a considerable body of program code to make it usable (eg, implementing consistency checks that could be specified relatively easily as a constraint with traditional modeling). That is, its use is analogous to major surgery—a resort to be employed in extremely limited circumstances. Further, the EAV plumbing is definitely not intended to be exposed to end-users: in other words, end-users should not be interacting with the EAV tables directly or even be aware of their existence.

By contrast, the traditional relational approach is easy enough to learn that, for small applications, even nonprogrammers can create their own microcomputer databases to meet basic needs, develop user interfaces to those databases, and work productively. As a result, the use of EAV is the *exception* rather than the rule, and it is *ancillary* to traditional approaches, being used only for those parts of the system where flexibility is mandated.

I'll return to the EAV approach when I discuss electronic health records and clinical study data management systems in chapter: Clinical Research Information Systems: Using Electronic Health Records for Research.

3.1.5.2 Schema-less databases

Some database technologies dispense with schemas altogether. This is the approach taken by several "NoSQL" database systems that are discussed in Section 3.5. These systems have the same flexibility as EAV systems. In fact, many of them model *all* data as Key–Value pairs (Key = Entity + Attribute): you can define new attributes and make many changes to attributes during runtime. Most of these changes do not require the software application to be stopped and restarted.

This approach is viable for systems where the database is accessed *only* by program code, not by end-user-oriented tools. Ultimately, however, as in the case of the EAV approach, there is no free lunch. Developers have to implement consistency checks (which are unavoidable) in program code. Also, a few changes to the system, such as implementing complex constraints, require changes to the program code. Therefore termination and restarting of the application is still necessary on occasion.

I believe that abandoning schemas *entirely* because a *part* of your application needs flexibility—unlike relational EAV modeling where it is *selectively* abandoned—is the equivalent of throwing out the baby with the bath water. The creators of some NoSQL systems that started out schema-less (such as Apache Cassandra) have apparently come to the same conclusion and have introduced schema-definition and data-manipulation languages in order to make developers more productive. These languages, in many cases, tend to be SQL-like to allow easier learning for developers who know SQL.

3.2 TRANSACTIONAL DATABASES VERSUS ANALYTICAL DATABASES

Databases serve two categories of functions.

1. To let users interactively *add, delete, or change* data. Such operations are called *transactions*.
2. To let users *analyze* the data, through queries, reports based on queries, or special analyses. The objective of entering all that data is to do something useful with it.

For databases with a small number of concurrent users, both functions can be served by the same database. As the number of users grows, this becomes increasingly difficult to do. Complex queries that return unpredictable amounts of data can cause transactions to slow down unacceptably. Throwing considerably more expensive hardware at the problem may mitigate the situation or it may not.

An alternative approach is to use *two physical databases on separate machines*. The first (*transactional*) database is used for moment-to-moment transactional operations. The contents of the second (*analytical*) database are refreshed in bulk from the contents of the transactional database, typically once every night, when the transactional system is relatively inactive. If the technology allows it, the refresh is *incremental* rather than full. That is, only *changes* to data are propagated: this can save time and machine resources for very voluminous data. Incremental updates are also desirable for systems accessed round the clock (eg, for industries operating in multiple shifts or when accessed from multiple geographical zones).

3.2.1 Transaction processing: basic concepts

In this section, I'll elaborate on transactions because one of the motivations for the recent development of nonrelational technologies is to be able to handle them in diverse situations where the data is distributed across hundreds to thousands of machines (such as with the datasets managed by Facebook and Twitter).

The characteristic of a transaction is that multiple operations that are part of the transaction are treated as a single, *atomic* operation. Consider an interbank money transfer. There are two operations: the first bank withdraws money from the account that is the transfer source, and then the second bank deposits the money into the destination account. The transaction is complete (*committed*) after the deposit completes. *Both withdrawal and deposit must complete, or neither must*—execution of only one step is unacceptable because it would result in the disappearance of money, or the spurious creation of money that doesn't exist. Any hardware, software, or network failures along any step of the way must result in a *rollback* (cancellation) of the entire transaction. Either after rollback or commit, the data contents must remain structurally sound, or *consistent*.

Ideally, transactions should be *isolated* from each other. That is, transactions from different users should not contend with each other or influence each other while they are still executing. This is because, until a transaction is actually completed or cancelled, we don't know whether the data changes that a given transaction makes will be permanent. In case two transactions need to access the same resource (eg, one transaction withdraws money from an account while the other adds funds to the same account), the transactions should appear to be *executed sequentially* (or *serialized*).

Also, once a transaction is executed, a record of it should not be lost later due to hardware failure. That is, transactions must be *durable*. If such loss were to occur, electronic commerce (even simple online bill pay) would be impossible because it could not be trusted. Durability is achieved through redundant hardware, as well as eventual offline (and offsite) storage.

The transaction processing system of all databases is designed so that certain operations are prevented under all circumstances, such as two people buying the same item. Thus, if an item is being modified by one transaction, it is prevented from being modified by a second transaction until the first transaction finishes, with either a commit or a rollback. In essence, transactions need to wait their turn.

The terms "atomic, consistent, isolated, durable" are known by the acronym ACID. The ACID approach is supported quite well by relational databases operating on structured data. For unstructured data such as documents, Web content, and multimedia, however, there are distinct needs related to the human workflow involved in their creation and modification, where relational databases currently provide little to no help. Take, as an analogy, the fact that two people can be drawing on the same picture interactively. A multimedia program that allows you to interact with another artist would be poorly served by a database because, by design, relational databases protect their data elements from being edited simultaneously by multiple users.

> *Note*: **The succeeding examples are drawn from nonmedical fields. This is because the traditional "bank-transaction"-type model continues to apply to hospitals and clinics where healthcare providers enter patient data. Also, the concurrency loads that hospitals face are orders of magnitude less than those faced by airline reservation systems, Wall Street, or Facebook: biomedical applications haven't really pushed the envelope here. But things may change, and so you, the reader, might as well prepare for the future.**

3.2.1.1 Isolation: easier said than done

Isolation is the hardest aspect of a transaction to achieve in practice. Complete isolation through serialization, which implies queuing of all transactions that require access to a particular unit of data, can cripple system functionality by increasing wait times to unacceptable levels. Fancy hardware will not solve this problem because very often the root cause is human. I explain next.

Consider the bank transaction described earlier. The *human* aspect of the transaction begins through a manual or electronics-assisted process—a customer approaches the bank physically or uses the website, etc. It continues with verification steps (eg, to ensure that the destination routing and account number is valid, that sufficient funds exist, and so on). The *electronic* aspect of the bank transaction, however (ie, the actual changing of data) takes only a couple of seconds, or even fractions of a second, if the transaction is within the same bank.

In many activities involving human workflow, however, transactions take minutes, hours, days, or even weeks: people take time to make up their minds and have the freedom to change them. They take considerably longer when creative work is involved. A few different scenarios are as follows:

- In an airline reservation context, when we are about to book a flight online, we may explore several alternative days and departures before choosing one flight, selecting a seat, and finally purchasing our ticket/s. Strict serialization would mean that when a passenger chooses a flight to look at available seats, all other passengers would be prevented from even *looking* at the same flight until the passenger is done, even though most passengers end up changing their minds and looking at alternative flights. It is as though they were queuing up at a virtual airline ticket counter, being served one at a time by a virtual salesperson.

  Also, consider how many shoppers dither over their choice of purchases. (A "Curb Your Enthusiasm" episode has Larry David getting into an argument with a particularly indecisive woman who is holding up the line in an ice cream store while sampling one flavor after another, and who he accuses of being a "serial sample abuser.") Software that forces you to wait in line, as telecom providers do over the phone when they ask you to wait for the "next available representative," can create a frustrating user experience.

- Software programs, magazines, and newspapers are created by teams of people. The same article may be reviewed by multiple people before being approved. Articles take time to get written and revised. If a strict queuing approach was enforced, only one person at a time could edit a document; others could not even look at its current state until this person was done.

Queuing may be unavoidable if you're waiting to be served by a limited human resource, such as a technical support person or a fast-food order taker. But in the context of transactions, it is possible to do much better if compromises are allowed that may occasionally frustrate a few but benefit the many.

3.2.1.2 *Limiting the degree of isolation to reduce wait times*

The simplest compromise is to allow others to look at items (ie, *query them*) while they are being edited. Since people usually look at *sets* of items (eg, when you are shopping on an e-commerce site), data that meets the criterion used to define your set may be changing under your very eyes through addition and deletion of new records by other users. Therefore, some of the items that previously met your criterion cease to do so (and vice versa). If the number of items is large, however, the fact that your set is slightly inaccurate does not matter most of the time.

This approach can also be used for the magazine/software scenario. Only one person can reserve an article or software module at a time for *editing*, while others can only *look* at it while accepting that this may not be the absolutely most recent version. (The act of reserving is called *check-out*, commit is called *check-in*, and rollback is called *reverting*.) It's not as powerful as allowing simultaneous editing—which is discussed later—but it's a first step.

The trickier situation is when someone is looking at data with an intention to change it (eg, make a purchase). For example, it is possible that, while you are looking at the available seats for a particularly busy flight, someone else may have just booked a seat that, on your screen, is currently shown as available. How do we avoid selling the same seat on the same flight to two different passengers?

The approach used here is called *optimistic concurrency*. It is called optimistic because it assumes that, most of the time, the user will not choose the item being viewed. (By contrast, serialization is a severe form of *pessimistic concurrency*: it assumes that a transaction will almost always be committed.) The idea is to have a field in the record within the item's data that serves as a kind of *version indicator*. It could be a number that keeps increasing by one each time a change is made, or it could be a time stamp, accurate to the fraction of a second.

Let's use the airline seat booking as an example. When you first decide to inspect the seats on a flight, the "items" are the seats and the version indicators are simply Yes/No values indicating that specific seats are free.

1. When you access the flight for the first time, the software grabs the version indicator value/s and stores them. You then take your time to make up your mind. If you decide to book a seat, a transaction begins. It reads the version indicator field *again* and compares it with the value that you have read previously.
2. If the values are the *same*, it means that nobody else changed this record (ie, booked the seat). The transaction then proceeds with a purchase and the status of that seat is changed. The transaction then commits. The entire electronic transaction executes within a fraction of a second.
3. If, however, the indicator values are *different*, it means somebody changed the record in the meantime. The transaction rolls back, the software issues an apology, explaining what happened, and takes appropriate action (eg, refreshing the screen so as to show you the present status of the flight's seats so that you can choose another). Sure, it's a little irritating, but it's better than waiting forever in line just to look at the seats in the first place. Also, unless there's a desperate stampede for tickets, the laws of probability favor scenario 2 over this one so the times you have to pick another seat are relatively uncommon.

3.2.1.3 Extending optimistic concurrency to documents

Software such as Google Docs and Microsoft Office 365 allows collaborative simultaneous editing of documents or spreadsheets over the Web. Different people can work on different parts of the same document and even see what others are doing while they are at it. This capability is useful in certain circumstances, such as collaborative editing of minutes during a meeting, brainstorming of ideas, or final tweaking of a newspaper/ magazine article. *To allow concurrent editing, the software treats the content as an aggregate of units.* In the case of spreadsheets, the unit is a cell. For a word processor, the unit is a paragraph, comprising one or more lines.

This approach has also been extended to "version control" tools that support collaborative programming, such as Git, created by Linus Torvalds, creator of the open-source Linux operating system, to support collaborative development and enhancement of Linux. The unit here is a line of program code. (Git is a *distributed* system where, instead of one single central "Master" copy of the program code, multiple copies reside in geographically scattered locations that are kept synchronized with each other as needed. Most of the time, each copy need not even be on a network: individual developers can work offline.)

Users working on different units do not interfere with each other, and it is straightforward for the software to merge the content as individual edits are committed. Such software will typically show you all the other users working on the same document and the changes each one is making in real time.

A few situations, however, complicate this simple picture and require manual resolution by a designated "authority" for that document. (The authority is designated beforehand.)

- If two or more users are making changes to the same unit (eg, the same paragraph), all changes are accepted, but only temporarily so. The software indicates a conflict and shows the alternative versions of the conflicted unit. The authority must now manually select which to use. Alternatively the authority could synthesize a new unit that combines the best aspects of individual edits.
- For program code and spreadsheets that contain formulas, there may be *dependencies* between different parts of the code—a change or error introduced in one place may cause a chain reaction of errors in other places. (In the context of program code, this is known as "dependency hell.") The problem may take time to troubleshoot, and the authority must find out which particular edit caused the problem. Once the offending edit is identified, there are two possibilities.
    - If the edit in question has an identifiable error, it must be rolled back.
    - If the edits turn out to be functionally correct, the code units or formulas that depend on them need to be modified appropriately to harmonize with the change.

(For those with programming knowledge: An example of a functionally correct change that causes dependent code to break is the modification of a subroutine by adding parameters.)

3.2.2 Transaction processing across distributed hardware

When the number of concurrent users is very large and geographically dispersed, as is the case with Amazon and Facebook, even a single unit of hardware may be inadequate for a transactional system. Instead, one has to employ several geographically scattered machines in a network. At this point, the complexity of transaction processing increases greatly, and the processing of transactions requires fine tuning to balance the following competing issues.

- Reasonable response time or *availability*: users cannot wait too long to access the system or make changes to data.
- Accuracy or *consistency*: changes to the data at one machine must propagate so that the user who made the change can see it and other users see those changes as well. In the distributed hardware situation, this will *not* happen instantaneously, because synchronization takes time.
- Resistance to system failure, also called *partition tolerance*: if a software system were to run on a single machine, any failures (due to hardware or software issues) would result in the system's being unavailable to all users until the problem was fixed. In other words, a single machine has *no* partition tolerance.

In distributed hardware, one designs the overall system to mitigate this possibility. The data, which may be voluminous, is *partitioned* (or *sharded*) across multiple machines, as well as *replicated* across machines. It is generally not worth replicating a data item across every one of the hundreds of machines in the system: change propagation would take too much time. Instead, replication is limited. A *replication factor* of 3 means that a given data item is stored on at least three machines: the machine where it originated and two others in the network. Each machine keeps track of the other machines where that item is stored, and this *location information* is propagated to other machines. This is the kind of setup Facebook uses.

If one machine fails, the user is invisibly redirected to one or other of the other machines where the data is replicated. These two machines do extra duty until the first machine is fixed. This way, we have achieved partition tolerance. To make this mode of operation feasible, the "item" must be a fairly large chunk of data so that updating only its location information on other machines takes many fewer networking resources than updating the data itself. This way, a user who needs to access that item from a fourth machine is invisibly redirected to one of the three machines where it is stored.

If all of this seems pretty complicated, that's because it is. Managing the necessary housekeeping is a challenge for software developers. Commercial RDBMSs, while supporting distributed setups, provide relatively modest help for distributed transactions—especially where (unlike bank transactions) changes such as updates to someone's Facebook page are permitted to be propagated in a relatively leisurely fashion. More important, buying hundreds of licenses for the distributed version of the RDBMS software can cost an arm, a leg, and a lung. This factor, more than anything else, has spawned the development of "NoSQL" databases: developers at ambitious start-ups (Google, Amazon, Facebook, etc.) decided that they were unwilling to be held hostage by another vendor and chose to build their own technology. Fortunately several of them were social minded enough to describe what they did, and even released a basic version of their technology as open-source. I discuss some of these technologies in Section 3.5.

3.2.2.1 The CAP theorem: trade-offs

A postulate originally proposed by Eric Brewer [7] called the *CAP* (Consistency–Availability–Partition tolerance) *theorem* states that in a distributed-hardware system (ie, one with partition tolerance), *one must trade off consistency versus availability*. If we want every machine in the network to always be immediately consistent with respect to the current data—so that as soon as data is added or changed, it is propagated to all machines before any user can proceed further—then response time would suffer. If we want reasonable response for all users, we have to postpone the consistency operation. The machines will *eventually* become consistent with each other, but not *immediately* so.

The catch lies in the word "eventually." How much delay is permissible depends on the nature of the application. If you make a change to your Facebook page's content, others don't need to see the change that very moment: a several minutes' delay would be acceptable. For bank transactions and airline reservations, the delay must be as short as possible: consistency (and synchronization with serialization) is paramount, even if it means that the overall throughput of transactions must slow down.

> **Without serialization of bank transactions, synchronization delay could be leveraged for a swindle: the example scenario given next is modified from Lloyd et al. [8]. A couple opens multiple joint bank accounts and deposit sums of money in each that fall below the ATM withdrawal threshold. A few days later, having synchronized their watches, they go to two geographically separate locations (eg, the US West Coast and East Coast) and at preagreed times, simultaneously empty every one of their bank accounts, relying on the delay in consistency to execute the same transaction twice against each bank, thereby doubling their money. (After this, one presumes that they immediately move to a country which has no extradition treaty with the USA.)**

Lloyd et al. point out that in many cases, *eventual consistency is not enough*, and that the system must exhibit *causal consistency* (ie, *changes should occur in the order that the user expects*). An example they provide is of a student who first removes photographs that show him drunk at a party from his Facebook page. He then sends his academic adviser a "friend" request. Assume that other machines in the Facebook network also have copies of this page. The two operations propagate to a remote machine (which the adviser, who is traveling, accesses) *out of sequence*, with an appreciable delay between the arrivals. The adviser gets the friend request first, opens the student's page (ie, views an out-of-date copy of the page), and sees the photographs. Causal consistency prevents this. Facebook implements "read-only" transactions: the adviser's action on the student's page (ie, looking at it) cannot commence until previous transactions on it (the content update, the friend request) have completed.

> **Another way to implement causal consistency is to use a variant on the way the Internet works. The Internet was specifically designed to withstand a nuclear attack that might sever direct point-to-point transmission. All machines are connected in a network to allow alternative routes between two geographically remote locations. When data is sent (in bundles called *packets*) to a remote machine, different packets may pick different routes to get to the same destination, based on momentary variations in traffic—just as we occasionally take detours from our usual commuting route.**

Packets may therefore arrive at a destination *out of sequence*. However, every packet has a *sequence number*, and the receiving machine's software uses this number to assemble them in the right order. Similarly, every transaction can be tagged with the page's current (sequentially increasing) version number. If the "friend request" version number doesn't match that of the remote page, the page's current version number on the remote machine, it means that the original page had changed in the meantime. So access to the copy is delayed until the remote page is synchronized with the original page.

3.3 DATABASE INDEXES

So far, I've talked about data (and its processing) scaling up to the size where multiple machines are required. I now discuss the underlying technology that makes such enormous scales possible while still providing adequate availability. *Indexes* are structures that make possible the efficient searching of large volumes of data. They may reside either on disk, in memory, or both. Without indexes, modern database technology would not be possible. Indexes can be used for all the types of data elements described previously. An electronic index serves the same purpose as a printed index at the back of a book. It saves you the trouble of scanning the entire content to locate an item of interest: consulting the index first, and then going to the page/s referred to, is much faster.

I will first introduce indexing of structured elements, which is simpler conceptually. To index semistructured or unstructured elements, one extracts *features* (which are structured elements) from them, and then builds an index on those features. Feature extraction is often a highly domain-specific problem: extracting features from a chest X-ray is quite different from extracting features from marked-up text or from an EKG tracing.

3.3.1 Uses of index-based search

Searching for specific values of structured elements (eg, finding details of a patient, given the patient ID) has other uses besides the obvious one. In systems where data is continually updated or new data is added, the use of indexes can enforce consistency through rapid implementations of actions such as the following:

1. When a particular value is entered, locate it and display associated data to make sure that the value was not wrongly entered. For example, display city and state when a zip code is entered, or product details when a Universal Product Code is entered or scanned. If the value does not exist (eg, an invalid product code was entered), reject the value.
2. Prevent inadvertent addition of duplicate values. For example, when a new patient record is created and SSN is entered, check it. If it already exists in the table, prevent saving of the record.

The design of indexing structures is a specialty within the field of computer science. Different types of indexes have been designed for different purposes based on the kind of data that you want to index, whether the index will reside mostly on disk or entirely in memory, and whether a single CPU or multiple CPUs can simultaneously access the index. The choice of index also depends on the kind of operations that you wish to optimize. For example, some index structures are preferred for tables that will be actively and continuously updated, while others are much more suited for data that will be updated only once a night and then not interactively updated at all, such as tables in analytical databases.

3.3.2 The B-tree family of indexes

The commonest kind of index, the *B-tree* family, is shown in Fig. 3.2.

The index is like an upside-down tree, with the root at the top and the branches and leaves below. In Fig. 3.2, each *Node* has two units of information: a last name (which we need to search) and a *record number* that indicates where (on disk) the record is located. Each node in the tree has two children: we say that it has a *branching factor* of 2. If you look at the names, you will find that for any node, the left child is alphabetically earlier (ie, "smaller") and the right child is later (larger).

3.3.2.1 Principles of tree-based search

Let us assume we want to locate the record number for Farrelly in the tree given earlier. Starting with the root node (Johnson), we see that Farrelly is *less than* Johnson, so we go to the *left* child (Charles). Now, since Farrelly is *greater than* Charles, we go to the *right* child and find Farrelly, whose record number is 6.

The process is analogous to searching a telephone directory for the phone number of a person with a known address. You can employ a technique known as *binary search,* a technique that can be used for any sorted data.

- You jump to the middle of the book. The top of the page typically has a three- to four-letter abbreviation of the names first and last phone numbers on that page.
- You now check whether the name of the person of interest is lexically before or after the name at the top of the page.

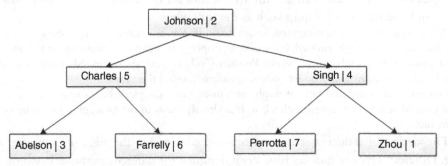

Figure 3.2 *A B-tree index.* Each node's content has two components: the field value and a record number, and has up to two "children."

- If the name is lexically *before*, it means that the person's number is in the *left* half of the book; if *greater*, it is in the *right* half. You can now repeat the procedure by jumping to the middle of the corresponding half of the remaining portion of the directory.
- You keep doing this repeatedly, *progressively bisecting the range of pages to be searched,* until you narrow down to a single page. You can now bisect the contents of the page until you finally locate the person (or fail to do so because they have an unlisted number).
- *With each step, the range of index content to be search shrinks to half the range of the previous step.* It can be shown that the number of steps required to locate a target (or not) is $\log_2$ (number of entries). This implies that if there are 1 million entries in the directory, it would take a maximum of 20 steps to locate a desired item.

Notice that when you traverse the tree of Fig. 3.2 from the root, going either left or right, you similarly reduce the number of items to be searched by half with each step. In other words, while a telephone directory is not laid out like a tree, the search principle is the same.

The tree structure is useful because when new nodes are added or deleted, there are techniques for ensuring that, by shuffling relatively few items around, the tree stays more or less *balanced*—that is, there is not much difference in the number of left descendants and right descendants from a node at any level of the tree, so that the number of steps required to search a tree of N items stays close to $\log_2 (N)$.

To allow insertion, we allow a node to temporarily hold up to two items, with the left item always smaller than the right. If, however, there is an attempt to add a third item to a node (because that is the only place it can fit in the tree), we *split* that node. With the split, the middle item becomes the parent, and the left and right items become the left and right children of the parent respectively. A nice animation (in Adobe Flash) that illustrates the continuous balancing of a tree as new items are added is at http://ats.oka.nu/b-tree/b-tree.html.

> **The B-tree was designed by Bayer and McCreight (both employees of Boeing) in 1973: the authors never stated what B stood for, but McCreight, in a lecture decades later, said that the letter B suggested the words Boeing, balanced, and Bayer (the senior author).**

3.3.2.2 B+ trees

The previous design makes sense only when the index fits entirely in memory. For indexes that are too big to fit into memory, 20 disk accesses for a search of 1 million items would be way too many. (Magnetic-disk access, which is limited by the time taken by the disk head to move to a particular location, is a million times slower than memory access.) To minimize the number of disk accesses, a modified structure called a *B+ tree* is used.

Here, each node may contain a large number of records (not just a maximum of two), as well as have a large number of child links. The maximum permissible number of child links (ie, the branching factor) is designated B. A node can have up to B−1 records. (If, during addition of records, the number is about to exceed this, the node is split, as in the

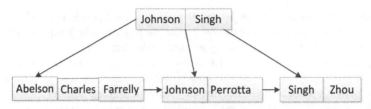

Figure 3.3 *A B+ tree showing the same data as Fig. 3.2, with a maximum branching factor of 4.*

B-tree of the previous section. An animation for the B+ tree is available at https://www. cs.usfca.edu/~galles/visualization/BPlusTree.html, which uses the term "Max Degree" for B. The animation is case sensitive, so make sure you use all caps, all lower case, or proper case.)

Fig. 3.3 shows the same data as Fig. 3.2, in a B+ tree with a value of B = 4. This tree has two levels. Note that there is additional capacity for another branch from the root node, as well as additional capacity in the middle and right leaves. (Fig. 3.3 is only by way of illustration. In real life, B is typically ≥100. Record numbers are not shown for reasons of space.)

As in the case of the B tree, "smaller" values are on the left and "bigger" values on the right. Note also that other than the leftmost leaf, the values of the record in the corresponding parents are replicated in the first record of the middle and right leaves. This way, the entire contents of the records at the "leaf" level form an ascending sequence. (The link between Farrelly and Johnson, and between Perrotta and Singh, indicate this.)

The value of B is determined by calculating the size of a node. This calculation is based on the size of the column that is being indexed plus space for the links. For example, integers take between 2 and 8 bytes depending on the largest number to be represented, date/times take 8 bytes, links take 8 bytes, and short text takes 1 byte per character + 2 bytes to track the number of characters. The value of B is chosen so that the node can just about fit on to a *disk block*. A disk block is a unit of disk space that can be read from or written to in a single disk operation.

> **In Microsoft Windows, the block size can be set to between 4 and 64 K when you first format a disk drive. The larger the block size, the faster the disk performance will be, at the cost of more wasted space when the size of the data being stored (a "file") is not an exact multiple of the block size. These days, disk space is cheap, so one can err toward the larger number in order to gain speed. Some setups, like Apache Hadoop, discussed later in Section 3.5, work best when your block size is very large (eg, 64 MB).**

The Wikipedia article [9] describes the B+ tree in depth. For now, it is worth remembering that a B+ tree requiring N levels (ie, a maximum of N disk reads) and a branching factor F can store approximately F^N records. Thus, if $F = 100$ and $N = 4$, the number of records is approximately 100 million. (The contents of a single node, which are sorted, are scanned by binary search in RAM.) With the basic B-tree, instead of 4 disk accesses, it would take $\log_2(10^8) = 26.5$ disk accesses.

3.3.2.3 Indexing trade-offs

In a database table comprising multiple fields, one can have B+-tree indexes on individual columns, as well as indexes on combinations of columns. For example, in the Visits table mentioned earlier, one could have a unique index for the Visit ID, and another unique index for the combination of Patient ID and date/time of visit. (A *unique index* is one that prevents duplicate values.)

There are drawbacks, however, to having too many indexes per table for tables that are continuously updated interactively. Every time a record is added to/deleted from a table, every index on that table has to be updated. This can cause the database to slow down when a large number of concurrent users are intensively entering data into the same table. Therefore one creates indexes only on those fields that need to be searched almost all the time (eg, in a table of patients, one might want to index by Patient ID, last name + first name, medical record number, and maybe date of birth). On the other hand, in analytical databases, where interactive updates are disallowed, one can index almost every single field if desired.

3.3.2.4 Uses of B-tree indexes

The B-tree index family is so widely used because it is "all-purpose."
- The index may contain duplicate values so that one can have an index on columns such as zip code. Alternatively, it can be a unique index, as mentioned earlier, to prevent duplicates.
- It gives reasonable performance when values are added, deleted, or changed, subject to the caveats of the previous section that the number of indexes on the table is modest.
- It is useful in supporting range searches, where one looks for all values in a certain range (eg, income between $20,000 and $35,000).
- For text, it allows efficient search of the entire text (eg, last name, such as "JONES," as well as prefix searches based on the first few letters, as indicated by "JON%." ("%" is the wildcard symbol.)
- B-tree technology can readily be adapted, with some preprocessing of the input, to deal with complex data. Approaches to indexing narrative text are discussed in Section 3.3.6, and semi-structured content in Section 3.3.7.

3.3.2.5 Multidimensional B-trees for spatial indexes

GISs are database applications that store and manipulate *spatial data*. Spatial data comprises objects that are either *points* or *shapes* in two- or three-dimensional space. Points have X, Y, and Z coordinates, which correspond to latitude, longitude, and elevation: the last may be ignored depending on the application. Shapes have *boundaries* within the same coordinate system. With modern global positioning system (GPS) technology, locations can be accurate to within a couple of meters. With augmentation technology, this improves to within a few centimeters.

Each object is associated with *attributes* that describe them (and values of those attributes). Thus, specific locations may indicate hospitals, schools, or grocery stores. The attributes associated with an object are highly variable and their number indeterminate. For example, if you zoom in on a restaurant in Google Maps, apart from its name, you may find that it serves Italian food, that it is moderately priced, that it has a rating of 3.8 out of 5 stars, and that it is open today between 11:00 am and 11:00 pm.

Examples of queries against systems, such as Google Maps, are "Find all (restaurants, movie theaters, etc.) within (specified distance) of my current location." Assume that one had separate indexes on the X and Y coordinates of all objects. One would answer this query through the following steps.

1. Find all objects within a certain range of latitudes.
2. Find all objects within a certain range of longitudes.
3. Find the intersection of the two sets obtained in the previous two steps.
4. Limit the list of objects by attribute and value of the attribute (eg, Business = "Restaurant," Cuisine = "Thai"). The limiting filter may be applied after Step 1 and Step 2, or after Step 3.

For a large database such as Google Maps, which records details on places all over the world—try "Thai Restaurants in Mumbai"—this approach would create two enormous sets in Steps 1 and 2 that would yield a relatively small final set after Step 3. It would be nice to be able to do *a range search in two or more dimensions simultaneously*.

This is what *spatial B-tree indexes* do. Think of a 2-D spatial B-tree as a tree in *three* dimensions. Here, a tree like Fig. 3.1 would represent coordinates in the X plane, and the Y plane would be represented by nodes sticking out of the page (toward you or away from you) at various heights or depths. Just as one goes left or right in a regular B-tree search step, here one would go left or right AND up or down. The gains in search speed are dramatic because with each phase of the search, one halves the number of search candidates in both X *and* Y dimensions. This would achieve average performance of $\log_2[\log_2(N)]$, where N is the number of objects in the map on which we have information. (For Google Maps, N is possibly in billions.)

3.3.2.6 Index fragmentation and reorganization

With continual insertions and deletions of large number of records over time (eg, several weeks to months), a B-tree index can get *fragmented*. That is, it ends up taking much more space than in an ideal configuration, and performance can suffer because of the extra disk accesses required. DBMSs therefore allow *index reorganization*, where the index can be rebuilt from scratch. Such rebuilding is done when the database is offline: it is a fast and efficient operation so that the duration of the offline period can be kept brief.

> **Certain database tables are themselves organized as B-trees: that is, a leaf node stores the entire record instead of merely the record number. Such tables are said to be *cluster organized*. (In Microsoft SQL Server, any table with a**

designated primary key is cluster organized by default, unless you specify otherwise.) Therefore, such tables are similarly vulnerable to fragmentation and may need to be reorganized periodically.

3.3.3 Hash indexes

While the logarithmic search performance of the B-tree indexes is impressive, there are other methods that work even faster—in more or less constant time. The *hash index* is often used for auto-number (primary key) table columns that represent unchanging, unique values that have no intrinsic meaning.

In a hash index [10], the number representing the primary key is transformed (using a mathematical function called a *hash function*) into another number called the *bucket address*—a location in memory or disk, also called a *bucket*, where that record is stored. It is possible that the mathematical function, instead of producing a unique address for a given key value, may produce the same address for more than one value (this is called a "hash collision"), and so a bucket is allowed to hold more than one record. Usually, the space allocated for the buckets is about 1.25 × (number of records) or more. With a well-chosen hash function, the number of collisions will be modest, so that very few buckets will hold more than two records. Note also that some buckets may be empty. Within a bucket, records are usually unsorted, requiring a linear search.

Hash indexes are used behind the scenes for operations such as automatic opening of doors almost instantly (or refusal to do so) in response to the swipe of an ID card with a magnetic stripe. They can grow to accommodate new values as records are added. They are, however, limited in functionality: they are intended to search a single value, not a range of values. Unlike B+ trees, however, they are unsuitable for nonunique values such as zip code. Since numerous people share the same zip code, numerous zip codes would hash to the same bucket: within a bucket, inefficient linear search would be required to locate an individual record.

3.3.4 Bitmap indexes

Bitmap indexes are useful when the number of distinct values that a column may take is "small" compared to the number of records in the table (ie, less than 1%). For example, in a table of US patients, there are only about 60 states + territories, only a limited number of genders (transgenders included) and a handful of races and ethnicities. For a database of a couple of million patients, even the set of distinct generic medications falls below the 1% threshold.

The idea behind bitmap indexes is that for each value, we create a *bitmap*, which is a sequence of 1s and 0s (ie, *bits*), each bit's position representing a record number. 1 means that the corresponding record contains that value, 0 means that it does not. The index for a column is the set of bitmaps for all distinct values (ie, if the column had only 10 distinct values, the index would hold 10 bitmaps).

Bitmaps are a very compact form of storage compared to B-trees because a single byte contains 8 bits, and modern CPUs typically handle 64 bits at a time. Therefore bitmaps can often readily fit entirely into memory: a million bits take up only 125 kB. Where there are more than 100 distinct values—this would be true of manufacturer + model in a vehicle database—it is obvious that most of the values in a single bitmap would be zero. In this case, one can achieve further space savings using compression techniques such as *run-length encoding* (eg, instead of recording a sequence of 98 zeros, just record the fact that zero has occurred 98 times).

3.3.4.1 Uses of bitmaps

Bitmaps are particularly suited for complex Boolean queries that combine multiple Boolean criteria based on distinct values and AND/OR/NOT operations. For example, we may want to find all patients who are female AND living in Texas AND who have ever been diagnosed with Systemic Lupus Erythematosus, but who have NOT received either of the drugs azathioprine OR belimumab. Boolean operations on bitmaps are supported directly in computer hardware. This is illustrated in Fig. 3.4. The AND operation (find patients who are females AND living in Texas) combines corresponding bits, so that the resulting bit, in the bottom row, is 1 only if *both* elements are 1, as for records 4 and 8: otherwise, the result is zero.

> **Historically, the 19th century British mathematician, George Boole, after whom Boolean operations are named, showed that if you treat 1 as *true* and 0 as *false*, "logical" operations can be defined on them, similar to addition or multiplication with numbers. Computer circuits ultimately operate on all data as 1s and 0s. Even addition of two numbers is implemented in circuitry as a sequence of logical operations.**

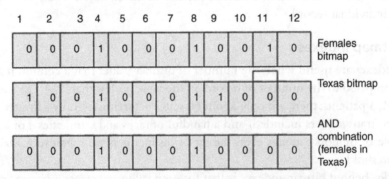

Figure 3.4 *The use of bitmaps to perform Boolean query.* AND operation: find patients who are females (top bitmap) AND living in Texas (middle bitmap). The bottom bitmap shows the result: it combines corresponding bits, so that the resulting bit is 1 only if *both* elements are 1, as for records 4 and 8; otherwise, the result is zero.

Compared to B-trees, bitmaps may or may not work efficiently for range queries: each item in the range must be treated as a separate bitmap. Thus, suppose we employed a bitmap index on year of birth. To find individuals born between 1950 and 1970, we would need to load and combine 21 individual bitmaps, one for each year in the range. Sometimes, one can "cheat" and create bins based on year ranges (eg, 1950–54, 1955–59, etc.). This approach works only for queries that ask for a range that matches exactly to the bin boundaries.

Bitmap indexes are not as useful as B-tree indexes in high-volume transactional databases. Unlike the B-tree balancing algorithms, approaches at maintaining the bitmap in synchrony with rapidly changing data are not very efficient. Bitmap indexes are more suitable for analytical databases whose contents are bulk-updated and then treated as read-only.

3.3.5 Alternatives to indexes: physical links

I now introduce a search method that is the fastest of all—one that eliminates indexes entirely. Suppose, given a foreign key (eg, a Patient ID in a Visits table), we wish to locate details of the record associated with the primary key (eg, details of the patient in the Demographics table). Instead of trying to locate a value in an index using B-tree search or hashing first, *we store, instead of the foreign key,* the *disk address* (the "physical link") *of the corresponding record in the primary-key table.* Then the software can take us directly to the place on disk where the desired details are stored. In other words, we use a direct hardware operation, instead of a preliminary search of the index. This can save considerable time when we need to retrieve the details of a large number of patients (eg, for use in a report or data extraction).

The difference between the two approaches can be understood by comparing a well-designed electronic book or a source such as Wikipedia, versus a printed book—though the software mechanism underlying hyperlinks is quite different. To locate a topic in a printed book (the index approach), you go to the index, search for the topic, and then turn to the page indicated. With an electronic book, you click on a hyperlink in the text and get taken to the page directly.

Historically, a type of database technology called *network databases* used physical links. Before relational databases, this technology was predominant. (The Saber airline reservation system used worldwide by travel agents—the Travelocity website was a spin-off—used network database technology.) Today, a technology called graph databases, which will be discussed shortly, has resurrected the physical link approach.

The physical link strategy has a significant drawback, however. *The advantage of indexes is that they provide a layer of indirection,* as explained next.
- Databases are never static in real life: their design keeps changing slightly to meet evolving business needs, and new tables occasionally get added that are linked to existing ones. This requires tables to be rebuilt and reorganized, which means that disk addresses of individual records change. (Reorganization of individual tables structured as B+-trees is also necessary at times to improve performance, as stated in Section 3.3.2.6.)

- When a table is reorganized, only the indexes on that table need to be rebuilt. Other tables that refer to this table don't need to be rebuilt, however, since they use only foreign key *values* and the values don't change.
- In contrast, with the physical link approach, rebuilding one table results in a chain reaction in which *every* table that referred to this one needs to be reorganized as well. Every table that refers to one or more of the altered tables needs reorganizing in turn, and so on. The resulting downtime may be unacceptable.

    In other words, *the link approach trades very high search speed for inflexibility*. This inflexibility was the reason why network databases lost out to relational technology. (In addition, network databases lacked a query language: it took days to weeks to write programs to extract data, while a nonprogrammer using SQL could achieve the same results in minutes. Further, the program code was highly vulnerable to the microlevel organization of the data: when this changed, the program code had to be revised. SQL, by contrast, is much more "high level," freeing the programmer to concentrate on the problem to be solved.)
- Graph databases are inflexible for the same reason. The links are constructed while the database is being loaded. It is expensive to change the links during runtime: therefore, for the very large datasets where they are used, such databases are typically reconstructed like analytical databases (eg, on a nightly basis). (In fact, graph databases *are* a special kind of analytical database.) This inflexibility limits their uses to niche applications such as analysis of social networks. The creators of many graph databases, however, have not forgotten the reasons for the demise of the network technology and have wisely included interactive (proprietary) query languages.

3.3.6 Indexing of narrative text

Large volumes of text need to be indexed in order to be searched rapidly, as is done by systems such as Google. The roots of the technology that Google employs were laid down in the 1970s by computer scientists such as Gerard Salton and Karen Sparck-Jones, who pioneered the field of (text) *information retrieval* as an off-shoot of the application of computing to library science. Here, individual records of text, such as scientific abstracts, manuscripts and even entire books, are termed *documents*. Documents do not necessarily lie within tables.

Manual indexing, which has been performed for at least the late 1500s, typically involves scanning a book page by page for keywords, and then creating a table that cross-references individual keywords with the page/s where they occur. Electronic methods of doing the same thing involve the following steps:

1. (Optional) Dividing the text into segments, such as chapters, sections (eg, title, abstract, body, paragraphs within sentences, and sentences within paragraphs). This step may be important in scientific manuscripts because a word occurring in the title or abstract can be given more weight than one occurring only in the body of the text.
2. Breaking the text into individual tokens (or *terms*). If Step 1 has been employed, one records against each term the details of its position in the document (sentence, paragraph, page, section, etc.). Tokenization of English text can use spaces and punctuation. For languages such as German, Hungarian, or Turkish, which string together individual phrases, prefixes, and suffixes to form compound words, the process is more involved. Tokenizing utilizes online dictionaries that specify how individual words are to be broken up.

3. (Optional) Very common words in the language—"the," "is," "from," etc.—may be removed because they have no value as keywords. Such words are called "stop words." Note that the more specialized the domain of the documents, the more the words that can be removed. Thus, "patient" would be a stop word in a collection of clinical notes.

   Google, which deals with arbitrary content, does not remove stop words at all, so phrases like "The Thing" (which could refer to the movie or the comic book superhero) get indexed.

4. (Optional) Words may be converted to a "root" form by eliminating inflections, using a table-driven process called *normalization*, or a more drastic rule-driven process called *stemming*. Thus, "swim," "swimming," "swum," and "swam" all get converted to "swim"; "go," "went," "gone," and "goes" become "go." Note that by doing this, the index becomes smaller, but specificity of search is reduced. Also, certain words are ambiguous. Thus, "left" can refer to the opposite of "right," or to the past participle of "leave."

   Google recognizes inflections, but does *not* eliminate the original words. (Try Googling "The Thing" and "The Things" and see the difference in hits.)

5. One consolidates the raw extracted information at multiple levels. (If Step 1 has been employed, one preserves the raw information.)

   a. The first step is to discard the position information (in case it has been extracted) and consolidate all occurrences of a given term in a document so that for each term–document pair, we get the number of times the term occurred.

   b. An additional step is to use this consolidated data and discard document information so that for each term, we get the number of documents that contain the term. As pointed out by Wilbur and Sirotkin [11], this step can be used to *discover* stop words in an unfamiliar domain: if the number of documents containing the word is very close to the total number of documents in the database, that word does not have very good discriminatory value as a keyword.

6. One creates a B+-tree index on both the raw extracted data (if Step 1 was employed) and on the aggregated data.

   a. The index on the position data, if created, is called the *proximity index*. It allows queries that ask for two terms occurring within the same sentence or paragraph, or within a particular number of words of each other. (Alternatively, the greater the closeness of the search terms in a particular document, the higher the weight that document can be given.) Note that many systems do not allow proximity searching because they do not create this index (which can be significantly larger than the documents that were indexed).

   b. The index on the aggregate created by Step 5a is called the *Term Frequency Index*. When the end-user specifies a search term in a query, documents in which that term occurs more frequently have more weight.

   c. The index on the aggregate created by Step 5b is called the *document frequency index*. When the user specifies multiple search terms within a query, the more uncommon terms are given more weight than the more common terms. Thus, in the query "acute meningitis," the latter term (an inflammation of the tissues covering the brain) is given more weight than "acute," which simply refers to a condition that developed quickly—and occurs so commonly in medical text as to almost be a stop word.

      The metric actually used for weighting is called TF × IDF (term frequency multiplied by inverse document frequency) and uses logarithms rather than the raw numbers. The metric also compensates for the document's length so that if a 300-word document and a 30,000-word document both contained a search term five times, the former would be given more weight, because the term is more likely to be a main theme within the former. Details are provided in the excellent textbook on IR by Manning, Schuetz, and Raghavan, and a quick overview in the Wikipedia article.

Indexes on text also support search using traditional Boolean operations. However, automatic "relevance ranking," as employed by Google (as well as bibliographic search engines such as PubMed) are the easiest to use, and therefore most widely employed by end-users.

There are various refinements to the indexing process.

- For scientific domains such as biomedicine, immediately after either Step 2 or Step 4 mentioned earlier, online thesauri may be employed, and *phrases* corresponding to phrases in the thesaurus (eg, "rheumatoid arthritis") are recognized and substituted if necessary with a root concept. Thus, "renal failure" and "kidney failure" mean the same thing; so do "vomiting" and "emesis."

    The process of concept identification can be quite challenging because of the presence of *homonyms* (phrases with multiple meanings) and ambiguous abbreviations. Thus, "cold" can refer to low temperature or to rhinitis, and DM can refer to the drug dextromethorphan (a cough suppressant used in cough syrups) or to diabetes mellitus. In principle, ambiguity can be resolved by determining context (the words before and after the term), but this is an elaborate process that is unsuited to bulk processing of text. It is often simpler just to flag the term as ambiguous and tag all its possible interpretations based on the thesaurus.

    The advantage of concept identification is *automatic query expansion*. Entering a search term and specifying that all synonyms also be looked for is an important convenience. (Enabling searching for all inflectional forms of a term serve a similar, but more limited, purpose.)

- Web search engines such as Google have several refinements that would not be necessary in, say, a database of scientific documents. Various web-page designers who are aware of TF × IDF will try to game the search engine by repeating a particular phrase (such as terms related to pornography or specific merchandise) endlessly in portions of the page that are invisible to the end-user, in order to increase their "relevance."

    Earlier web-search engines such as the now-deceased AltaVista were fooled completely, so that the search results were often pure garbage. Google has algorithms that recognize such repetitions, so that such pages and/or the websites that try to employ these tricks are simply blacklisted.

- Google also uses a technique called PageRank, where a particular web page is weighted based on the other pages that link to it. Because of the phenomenon of "Google Bombing," where a clique of pages can link to each other, there are various refinements to prevent PageRank-based gaming, typically as part of an elaborate prank. (At one time, typing "miserable failure" into Google would bring up the official White House biography of George W. Bush as the topmost hit.)

3.3.7 Indexing of semistructured data

There are two aspects of indexing XML or JSON.

- The documents have a regular structure determined by the tags or curly braces: one can extract individual elements within tags and associate them with their enclosing tags and create B+ tree indexes on the extracted data. There is a SQL-like language called XQuery

(XML Query Language) that can utilize the indexes. (Just like SQL, however, XQuery is supposed to work, albeit much more slowly, even if indexes have not been created.)

- In addition, if the individual elements themselves have arbitrary size, as in web pages, one can discard the tag information, treat the remaining text context just like unstructured text, and index it using the approach of the previous section. This is what Google does with web pages: RDBMSs index PDF or MS Word files (which contain embedded XML) similarly.

3.4 MANAGING INTEGRATED (STRUCTURED + UNSTRUCTURED) DATA

Organization of data into tables with rows and columns is the basis of relational databases, as I've discussed. In principle, however, there is no reason why a column could not contain semistructured or unstructured data, and most relational databases do allow this.

When RDBMSs were first introduced, they provided no support for managing arbitrary-sized data. Later, RDBMSs did little besides let you store and fetch such data as "Binary Large Objects": you could not query their content using SQL, and any manipulation had to be done using application program code. These limitations made such data hard to work with. Consequently, software vendors unassociated with RDBMS technology arose to fulfill this need, and many became commercially successful. As technologies such as XML and narrative text became mainstream—the former as a medium of data interchange, and the latter due to technologies that searched text, such as Google—the RDBMS vendors were forced to catch up. However, the facilities for manipulating such data depend on the particular RBDMS that you are using, for reasons now discussed.

The ISO SQL committee periodically issues enhancements to the SQL language specification to meet emerging needs. The functioning of this committee, however, has been characterized by bureaucratic and political wrangling originating in intervendor tussles, and the committee itself convenes infrequently because its members have day jobs. As a result, when the committee finally comes to an agreement, several years have elapsed. The new specification is often years behind the state of the art because the non-RDBMS software companies who originated the innovations that SQL is copying have not exactly been sitting still.

Individual RDBMS vendors, who have to meet their customers' needs, get tired of waiting for agreement and implement their own version of the necessary feature set. As a result, SQL has never been a single language, but a family of dialects with subtle differences between vendor versions. Oracle and Microsoft SQL Server, for example, differ in the facilities available to handle date/time data, or in the way a new table can be created from the results of a query. The divergence with respect to unstructured or semistructured data is far greater. (Eventually, for ISO SQL compatibility, the vendor may also support the "official" syntax, but this support is typically delayed unless specific ISO recommendations are so elegant that their benefits are immediately perceived by all.)

3.4.1 Choice of toolset for managing mixed data

When managing mixed data, you can use the full-text indexing and search capabilities of the RDBMS that you are using, or you can employ a dedicated package that can accept data exported from the database and index it. An example of a dedicated package is Apache Solr, which is intended to index and query text, is free, and is built on top of the respected Lucene technology. A commercial (but extremely affordable) option on the Microsoft Windows platform is dtSearch™.

- The advantage of using the native database facilities is the ease of integration with the structured data within a database. Many queries you ask of the data will be *mixed*, in that they will specify criteria based on the structured data in the database as well as the un/semistructured content. The RDBMS's query optimizer will typically use statistics on the indexes on both the structured data and the unstructured data to most appropriately narrow down to the appropriate hits.

   Thus, if the keywords specified in the query against the text data are highly selective, it will use the unstructured data as a filter. If the criteria against the structured data are more selective, it will employ those first. (In SQL Server 2005, this was not the case: the text search was not integrated with the database's query optimizer [12]. Microsoft SQL Server fixed this problem with SQL Server 2008.)

   An additional advantage of using the RDBMS is relative ease of setup, which does not require programmer-level expertise—only administrator-level access. In essence, you just point the indexing engine at the tables you want to index, identify the fields you want to index, and start the process. The fields that are candidates for text indexing typically are the fields containing lengthy text: short-text fields such as Last Name are usually best treated as structured data. Whenever records in these tables are added, deleted or changed, the indexes will be maintained. However, ease of use is not limited to RDBMS. While some of the free UNIX packages definitely require some developer chops, dtSearch on Windows is remarkably easy to set up.

- The disadvantage of using the RDBMS exclusively is that you sacrifice some customizability and search power. Solr and dtSearch, for example, support fuzzy search, which allows and corrects for limited misspellings in the search terms, as well as phonetic search (sounds like): SQL Server does not. Similarly, dtSearch supports *keyword-in-context display*, where a nutshell of each hit (eg, the sentence containing the hit, and the sentences before and after) can be displayed with the matched words highlighted. This is not too difficult to program, but it is nice to have such a feature out of the box. Similarly, dtSearch will automatically "crawl" across a specified list of external websites periodically to gather and index content. With an RDBMS, you have to first download the web pages, and implement or purchase your own web-crawler software.

   Another advantage of dedicated packages is that many of them (such as Lucene and dtSearch) are highly customizable with respect to the relevance ranking algorithm, and with respect to what tokens are recognized as punctuations versus characters to be indexed. (Thus, Lucene treats the underscore character by default as a word breaker/punctuation; dtSearch, by default, treats it as part of the word.) RDBMSs, by comparison, are black boxes whose internal behavior may not be fully documented.

My recommendation is that, unless you know from the beginning that you crave the advanced features, start with the RDBMS solution first. You can always move to the dedicated-package option later if you need to.

3.5 NONRELATIONAL APPROACHES TO DATA MANAGEMENT: "NOSQL" SYSTEMS

I now move on to alternatives to RDBMSs, which are not as suitable as general-purpose solutions for managing a variety of data, but which have evolved to meet very specific needs.

The term "NoSQL" refers to a family of database technologies created relatively recently that employ a nonrelational approach to database organization. "NoSQL," which originally stood for "not only SQL," is actually a meaningless marketing-type term. Many of the individual technologies share no common feature other than a nonrelational approach and an orientation toward supporting "application databases." The latter term, defined by Pramod Sadalage and Martin Fowler [13], refers to databases that are controlled and accessed by a *single* application, as opposed to "integration databases," which are accessed by a variety of applications as well as by power users who use interactive query languages.

There are several distinct motivations for departing from the relational approach.

1. *Dealing with unstructured or semistructured data*: RDBMSs were laggards in addressing needs related to these, as mentioned in Section 3.4. Today, they have mostly caught up, and so this is less of a motivation by itself (if the data volumes are modest eg, <100 GB) for using a NoSQL solution.

2. *Modified transaction processing with distributed hardware*: As discussed in Section 3.2.2, RDBMSs don't scale well to highly distributed hardware, especially when the requirements dictate a different transactional model than the ones that RDBMSs support well.

3. *Parallel processing with distributed hardware*: Certain applications (such as searching extremely large databases or processing of a vast amount of data) involve a high degree of *parallelism*. That is, a single conceptual operation is run simultaneously (ie, in parallel) across hundreds or thousands of machines, each of which holds a slice of the data. The results are then collated. Google operates this way (both for executing searches as well as for indexing web content), as does NCBI's BLAST search for biological sequence similarity. This programming approach is called *MapReduce (map* implies distribution of the task, *reduce* implies collation of the results). "Big Data" analytics rely on MapReduce greatly.

    Division of data—also called *sharding*, a term introduced in Section 3.2.2—is always accompanied by replication to allow for the possibility of failure or unavailability of individual machines. Sharding and replication are supported transparently by certain NoSQL systems, notably Apache Cassandra [14] (discussed shortly), which readily integrate with the most widely used map-reduce framework, Apache Hadoop [15].

    RDBMSs have begun taking baby steps in this matter: Microsoft SQL Server Integration Services can also access Hadoop-based data services. Programming MapReduce using Hadoop with traditional languages can be moderately tedious: a high-level programming and infrastructure framework called Apache Pig allows the succinct expression of data-analysis programs that must utilize parallelism.

4. *Special analytical needs*: These are discussed shortly in the context of graph databases.

5. *Flexibility*: For the relatively few occasions where the data elements are highly variable, some developers have opted for nonrelational approaches. I don't believe that flexibility alone is a sufficiently good reason to depart from relational approaches, unless one of the other reasons

given earlier also applies to a given problem. "Flexibility" (alternatively described as "we don't need no stinkin' schema") sounds great until you discover the amount of code you have to write to create an infrastructure to ensure that garbage doesn't creep into your data, and to make the system operable by end-users.

While cost considerations played an ancillary role in the development of some NoSQL systems, cost is not a driving factor *by itself* for a NoSQL solution. Robust open-source RD-BMSs such as MySQL and PostgreSQL are available as free (albeit unsupported) versions. Given the vast number of developers who are comfortable with relational technology, the opportunity costs of the time spent acquiring expertise with an unfamiliar DBMS, to the extent where they can build industrial-strength applications, should not be underestimated. The developers must still be paid while they spend their time learning. (Even altruistically minded programmers rarely come free other than for occasional small chores: they still need to pay their bills and support their families.)

3.5.1 Types of NoSQL systems

The term "NoSQL" encompasses quite diverse underlying designs. Currently, the following broad categories exist: more categories may arise, because DBMS technology is evolving rapidly.

1. *Key–Value stores.* More accurately, these are key–value–timestamp triples: the timestamp, which records the date/time of last modification, is used to support optimistic concurrency, described in Section 3.2.1.2. Some systems existed long before the term "NoSQL" was coined. The value may be arbitrarily large in size.

    a. In the earlier systems, the DBMS merely takes care of storing and fetching the data: manipulating the values is the responsibility of an application program. A well-known key–value DBMS, BerkeleyDB, served for a long time as the back end of the open-source Subversion version-control repository software, which is still used by small to medium-sized teams to manage their software projects. Currently, I can think of no reason to use such systems because the functionality is so basic, and the programmer has to do too much work. Even Subversion moved away, in 2005, to an alternative storage mechanism.

    b. The modern systems support manipulation of the value, which can be a simple (structured) or unstructured data type, depending on the service. The most basic type of "cloud storage" provided by Amazon, Google, and Microsoft (Azure Table Storage) is of this type. These systems are useful for storing multiple terabytes of data relatively cheaply: the data is transparently divided across multiple machines when necessary, and is also protected against loss by being replicated across data centers.

Cloud computing is a mechanism by which powerful CPUs and/or very large units of storage hardware, and the software to manage your data, can be rented and accessed over the Internet rather than purchased. The ideas are that:

- **The chores of machine administration, backup, and so on are taken off your hands.**
- **The rental operates on a "pay as you go" basis, so that for short-term projects or pilot experiments that require enormous storage, renting works out much cheaper than purchase.**
- **You can increase or decrease the resources rented based on your needs. In most cases, the cloud service can scale as needed dynamically; thus**

providing similar performance metrics as the number of users hitting the system grows.

- Cloud rental also comes with some guarantee of service: you do not generally have to be concerned about where the data is physically stored. Through replication across multiple data centers, if one data center is offline, you are invisibly directed to another behind the scenes.

Cloud services are not a panacea. There are some minefields: many cloud services have had teething problems and well-publicized service outages, so the service guarantee may not be worth much. Also, as I state in a future chapter, the US National Security Agency has back doors into several cloud services, so intellectual property may not entirely be safe from prying eyes. Further, as pointed out by Goodman [16], many of the providers of free storage (notably Google) once wrote their user agreements in a Byzantine legalese that appeared to give them the rights to copy and reproduce your work without paying you a dime: in other words, the "free" service wasn't really free.

> Modern key–value systems are useful, with the caveat that some of them may be too basic for many data needs. For example, Microsoft's Azure Data Storage only supports simple (structured) data types, albeit with a limit on storage of 100 TB per database. Microsoft also offers a relational database in the cloud (SQL Server Azure). This was offered after developers accustomed to RDBMSs found themselves traumatized by being forced to write code endlessly to do things with Azure Table Storage that RBDMSs do out of the box. The downside is that SQL Server Azure is limited currently to 150 GB per database (about 1/700th the size of Azure Table Storage).

c. DBMSs such as InterSystems Caché, which dates back to the 1990s, support arbitrary hierarchies. That is, a value can itself be a set of key–value pairs, and so on. Caché essentially resurrects the ideas behind the venerable MUMPS technology developed at Massachusetts General Hospital in the 1960s for the development of medical software. Despite the fact that Epic, currently USA's biggest vendor of EHRs, uses Caché as its engine, as does the US Veterans Administration Medical Centers' EHR, I can think of little reason to employ it; too much wheel reinvention becomes necessary.

The choices of Epic and VAMC were mandated because their systems were originally implemented in MUMPS, a technology that was already aged by the early 1990s, and Caché provided a migration path to more modern technology and hardware. However, both organizations had to build their own schema managers and a data-type system, in the fashion of RDBMSs, *on top of* Caché. If you don't have legacy MUMPS applications to migrate, I see no rationale for employing this proprietary technology, which is falling further and further behind both RDBMSs as well as open-source NoSQL systems that have vibrant communities. Some criticisms of Caché are particularly vehement [17]. It is hard to find talented programmers who would voluntarily want to work with it.

2. *Document-oriented DBMSs* represent an evolution of key–value stores, where the values are of a specialized type, typically XML or JSON content (Fig. 3.1). They allow manipulating the contents of values at the DBMS level, and facilitate search of value content through B-tree-based indexing, described in Section 3.4. They also support sharding for large datasets.

Document-oriented DBMSs are useful for blogging platforms (blogs are documents, after all) as well as specialized real-time analytics applications. The schema-less nature of such systems allows new data models to be simulated without actually changing schemas. Similarly, many e-commerce applications involve the use of schema-less approaches to model accurately the hundreds of different categories of products that are being traded, with minimal downtime. A popular open-source system, MongoDB, uses a binary version of JSON for efficient storage and indexing, and is finding a large number of uses: the popular Trello project-management software (available in both free and commercial versions) uses MongoDB as its back end.

3. *Graph databases,* mentioned briefly in Section 3.3.5, are tailor made for domains like social network analysis and recommendation engines. Recommendation software shows you, when you view or after you purchase an item, the other items chosen (viewed or bought) by people who chose this item. On RDBMS technology, such queries take a very long time to run.

   These are admittedly niche applications. These systems are schema-less: however, to facilitate use, some of them employ interactive query languages. The most popular of these, Neo4J, described in the book by Robinson et al., is available as freeware as well as a commercially supported, high-end version that can run on a cluster of machines. Neo4J's query language, Cypher, is SQL-like.

4. *Column-family stores*: The most well-known of these is Apache Cassandra (originally developed by Facebook). A column family in Cassandra (which looks to the user like a table) has rows. Each row contains a primary key (of one or more fields), and an associated hash table consisting of an arbitrary number of attribute–value–timestamp triples, all tied to that particular key value. The primary key is used to order the rows. *The attributes appear to the user as columns: however, for a given row, certain attribute–value pairs may not exist.* The timestamp, which is set when the column's value is initialized, is used to implement optimistic concurrency. It can also be set with a time limit for expiration, which means that the column and its value will automatically be deleted after that limit has lapsed, unless the time limit is programmatically reset in the interim.

   Cassandra uses the Cassandra Query Language (CQL) for schema definition and data manipulation. CQL has been designed deliberately to look like a subset of SQL. New attributes/columns can be defined at runtime, and the values may be complex, user-defined data types in addition to simple, structured types.

   Cassandra is probably the NoSQL database with the best support for entity–attribute–value (EAV) modeling (Section 3.1.5.1), which is used extensively in biomedical applications to model sparse clinical parameters. In addition, Cassandra can be put to the same uses as document databases, as well as website-related chores like page-visit counters and website shopping carts. The expiring timestamp feature can be used to easily implement user sessions on websites, which expire if the site has not been accessed after a certain time. Thus, if a user who is shopping on an e-commerce website leaves without purchasing anything, the user's data/shopping cart can be cleaned up automatically.

   Amara Healthcare Analytics uses DataStax (a commercially supported version of Cassandra, with enhancements such as integration with Apache Solr) for their Sepsis alert system, which is provided to hospitals either as a standalone solution, or as a SaaS (software as a service) over the Internet. The system receives real-time inputs from multiple sources: laboratory data, medical diagnosis codes; clinical narrative and monitoring devices [18].

3.5.1.1 Columnar-store implementation in RDBMSs

Columnar-data stores for sparse data are now considered such a success that RDBMSs (specifically Microsoft SQL Server 2012 and later) have appropriated the idea in a modified form for use in analytical databases where the tables are mostly read-only. Microsoft calls these "column-store indexes," though the data structures employed are not indexes in the strict sense of the term. Unlike indexes, which are *ancillary* to the records in a table, column-stores *usually represent the data itself.*

Here, the entire table is "shredded" into its individual columns, as shown in the right half of Fig. 3.5. After shredding, the record number for each value is not stored: instead, the *position of a particular element represents its record number* (in a manner similar to bitmaps; Section 3.3.4). Shredding is particularly useful for tables that contain a very large number of columns (as in seen in certain tables in analytical databases.) The idea of shredding is twofold:

1. Many queries on analytical databases ask for only a handful of columns, but operate on all values of those columns. If the data were record-organized, the DBMS would end up reading all of the data row by row, only to discard most of it. With column-organized data, the DBMS can retrieve only those shreds, and join them side by side to return the answers. (This assumes, of course, that record numbers do not change, which is true for analytical databases.) Consequently, there are fewer disk reads required to access the required data.

2. Certain columns (as for bitmaps) have a relatively small number of distinct values. Therefore, compression techniques (as for bitmaps) can be employed to achieve dramatic reductions in space requirements. (Sparse data, where many columns have no data, is also compressed efficiently.) Therefore, many column-stores may fit in memory where the data organized by row might not: even if they don't, fewer disk accesses are needed. This can greatly improve query performance.

In SQL Server 2016, Microsoft allows column-store indexes even on tables that are actively updated, but that still need to be queried for real-time reporting. The idea is that, in transactional databases such as those used in e-commerce, changes to records that have already been created are relatively few (even though new records are being added continually). Therefore, the DBMS keeps a record of changed, deleted, or newly created data in a separate data structure called a delta-store ("delta" is the Greek symbol used in

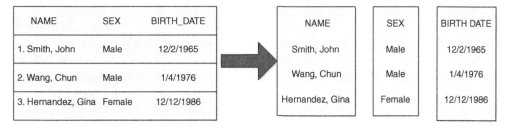

Figure 3.5 *Illustration of the "column-store" table design.* A table is "shredded" into individual columns; the shredding allows opportunities for significant compression, especially for columns that contain limited sets or ranges of values.

mathematics to denote changes to a variable). At periodic intervals, when the delta-store's size exceeds a particular limit, the column-stores are rebuilt by merging the new records.

3.5.2 Overview of uses for NoSQL databases

It is very uncommon, in biomedical databases (as well as in business applications), to find NoSQL solutions being used exclusively. Their use normally tends to coexist with the use of RDBMSs for the very same data. Sadalage and Fowler [13] clarify this issue, pointing to an idea of their colleague, Neal Ford, termed "polyglot programming" [19]. ("Polyglot" = one who speaks multiple languages.). This term simply means that different developers and analysts on the same team will do different things with the same conceptual data, using a variety of tools—and specifically, will employ the right tool for the right job.

Individual tools typically require data to be in different formats, and therefore it makes sense to have a single source for the data—a so-called *integration database*—from which subsets can be transformed into separate problem-specific *applications databases*. Each application's database will employ format appropriate for individual tasks. RDBMSs currently have a significant advantage in that using either SQL or third-party tools, such transformations into application-specific are relatively simple because of the conceptual simplicity of the relational model itself. Each applications database may possibly employ a different NoSQL DBMS.

An example of the polyglot approach in the biomedical area is data originating from the EHR.

- In some systems, such as Epic™, the production database, specialized for transactions, employs nonrelational technology.
- The analytical database, generated nightly from the production system, employs an RDBMS.
- If a researcher were investigating a specific set of epidemiological problems (eg, given a patient with one or more disease conditions, list the patients who live within a certain radius of this patient who suffer from the same diseases), one might import a subset of the query database into a graph database.
- A dataset of clinical notes pulled from the query store and marked up with XML as part of an NLP-processing effort on the raw data, might employ an XML-oriented document database or a special limited-purpose RDBMS where the XML is indexed.
- If de-identified data needs to be shared with a consortium of other institutions, the shared repository may well employ a key–value store that is hosted with a cloud provider.

NoSQL DBMSs seem to be adapting to coexisting in the RDBMS world. Where they do not directly implement a SQL-like language, many of them provide an Open Database Connectivity (ODBC) [20] interface for developers to utilize. (ODBC is an open standard originally developed by Microsoft in 1992 as a vendor-independent framework, which allows application developers to write a single body of code that can operate against a variety of databases. *It accesses a data-source via SQL, irrespective of the actual technology employed by the database.* The software that translates ODBC SQL to vendor-specific operations is provided by the vendor or by third parties.)

One warning: currently, with NoSQL technologies, you're on the cutting edge, which also means the bleeding edge. There are quite a few pitfalls hiding in the lurch. I discuss some of the security concerns in a later chapter. One well-publicized crash of the infamous Healthcare.gov was traced to the contractors' unfamiliarity with MarkLogic, a commercial NoSQL XML database [21].

Most of the problems, to be fair, were traceable to management failures by Department of Health and Human Services personnel, who lacked a sense of urgency, kept changing system requirements, and failed to fix the cumbersome processes in place. Many of the staff on the programming teams weren't exactly the best and brightest either, so rather than blaming the tools, one may have to blame the folks who didn't know how to use them. However, management should have calibrated the choice of toolset to the background and expertise of their staff, rather than toss their programmers into deep water without life vests on a mission-critical project. You can read the blow-by-blow account on Wikipedia [22].

As with all new technologies, you should probably employ it for pilot projects first and get familiar before you entrust a production operation to it.

3.6 FINAL WORDS

I've spent so much space describing storage technology for two reasons.
1. Informatics deals with data, and therefore basic database knowledge is essential. Polyglot development ability is increasingly likely to be part of your job description, and so it is worth being acquainted with the spectrum of what is out there.
2. Despite the diversity of technologies, certain basic concepts, such as primary keys, database constraints, indexes, and the concepts related to transactions, apply to all of them.

BIBLIOGRAPHY

[1] World Wide Web Consortium. Extended markup language (XML). Available from: http://www.w3.org/XML/, 2015.
[2] Introducing JSON. Available from: http://www.json.org/, 2013.
[3] E. Codd, A relational model for large shared data banks, Commun. ACM 13 (6) (1970) 377–387.
[4] J. Melton, A.R. Simon, J. Gray, SQL 1999: Understanding Relational Language Components, Morgan Kaufman, San Mateo, CA, (2001).
[5] P.M. Nadkarni, Metadata-Driven Software Systems in Biomedicine, Springer, London, UK, (2011).
[6] V. Dinu, P. Nadkarni, Guidelines for the effective use of entity–attribute–value modeling for biomedical databases, Int. J. Med. Inform. 76 (11–12) (2007) 769–779.
[7] E. Brewer, CAP 12 years later: how the rules have changed, IEEE Comput. 45 (2) (2012) 23–29.
[8] W. Lloyd, MJ. Freedman, M. Kaminsky, DG. Andersen. Don't settle for eventual consistency: stronger properties for low-latency geo-replicated storage. ACM Queue 2014;12(3).
[9] Wikipedia. B+ tree. Available from: http://en.wikipedia.org/wiki/B%2B_tree, 2010.
[10] Wikipedia. Hash table. Available from: http://en.wikipedia.org/wiki/hash_table, 2010.
[11] W.J. Wilbur, K. Sirotkin, Automatic identification of stop words, J. Inform. Sci. 18 (1) (1992) 45–55.
[12] J. Fisk, P. Mutalik, F. Levin, J. Erdos, C. Taylor, P. Nadkarni, Integrating query of relational and textual data in relational databases: a case study, JAMIA 10 (1) (2003) 21–38.

[13] P. Sadalage, M. Fowler, NoSQL Distilled: A Brief Guide to the Emerging World of Polyglot Persistence, Addison-Wesley, Reading, MA, (2012).

[14] N. Neeraj, Mastering Apache Cassandra, 2nd ed., Packt Publishing, Birmingham, UK, (2015).

[15] T. White, Hadoop: The Definitive Guide, 3rd and 4th ed., O'Reilly Books, Sebastopol, CA, (2015).

[16] M. Goodman, Future Crimes, Doubleday, New York, NY, (2015) p. 243.

[17] morbiuswilters. Intersystems Caché—Gateway to hell, 2014.

[18] DataStax Corporation. Case studies: Amara Healthcare Analytics: Amara Health uses DataStax to proactively protect patient's health. Available from: http://www.datastax.com/wp-content/themes/datastax-2014-08/images/case-studies/DataStax-CS-Amara.pdf, 2015.

[19] N. Ford, Polyglot programming. Meme Agora. Available from: http://memeagora.blogspot.com/2006/12/polyglot-programming.html, 2006.

[20] K. North, Multidatabase APIs and ODBC, DBMS 7 (3) (1994) 44–59.

[21] E. Lipton, I. Austen, S. LaFraniere. Tension and flaws before health website crash. New York Times. November 22, 2013. Available from: http://www.nytimes.com/2013/11/23/us/politics/tension-and-woes-before-health-website-crash.html, 2013.

[22] Wikipedia. Healthcare.Gov. Available from: https://en.wikipedia.org/wiki/HealthCare.gov, 2015.

CHAPTER 4

Core Technologies: Machine Learning and Natural Language Processing

4.1 INTRODUCTION TO MACHINE LEARNING[a]

Portions of this chapter may be elementary to some readers and may be skimmed.

Machine learning (ML) is a field at the intersection of computing and multivariate statistics. It was so named because it involves the design and implementation of computer programs that detect patterns in data (ie, they appear to "learn" from data). Some modern authors prefer the term "statistical learning," which is more accurately descriptive: I'll elaborate on this shortly. The list of applications where ML is used is vast: it includes business and finance applications (fraud detection, stock-market analysis, detection of associations between purchases of different items), epidemiology, computer gaming (eg, chess), robotics, information retrieval, e-mail spam filtering, natural language processing, pattern recognition (eg, signal processing, speech and handwriting recognition), and analysis of biological sequences (eg, identifying features that suggest specific functions in DNA or proteins).

While machine learning began as a branch of artificial intelligence (AI), it has since drifted away, emphasizing statistical foundations, while classical AI focused on manipulation of symbols rather than numbers. This is not to say that symbols are unimportant: the most powerful approaches combine both as needed, without being dogmatically tied to one or the other.

4.2 THE BRIDGE BETWEEN TRADITIONAL STATISTICS AND MACHINE LEARNING

ML research has impacted traditional statistics by more systematically dealing with intrinsic variability within a data set. I introduce the connection between machine learning and traditional statistics through the following example, which also introduces some basic ML terminology.

Linear regression is one of the most widely employed statistical techniques: even non-statistical packages such as Microsoft Excel provide it. Here, one has a number of *input*

---

[a] Files to refer to: Introduction to Machine Learning, Andrew Ng's CS 229 course, Data Science for Business.

Clinical Research Computing. http://dx.doi.org/10.1016/B978-0-12-803130-8.00004-X

variables, X_1, X_2, X_3 ... and wishes to predict, from a collection of data, the value of an *output variable, Y,* on the assumption that the data fits the equation:

$$Y = \alpha + \beta_1 X_1 + \beta_2 X_2 + \beta_3 X_3 \ldots \qquad (1)$$

where α and the individual βs are constants whose values must be estimated from the data. (This is done using matrix algebra techniques implemented in computer software.) The Xs are also called *independent variables, predictors,* or (in machine learning) *features,* while Y is also called the *dependent variable, outcome variable,* or *response.* In the simplest case, there is only one X, so that we have two-variable regression. The term "linear" is employed because in the two-variable case, one tries to find the straight line that fits the data best: when generalized to multiple dimensions, one thinks of a "hyperplane" instead of a line.

Of course, the hyperplane of best fit will not align perfectly with the data: there will be a scatter around it, as shown in Fig. 4.1, and the line is chosen such that the scatter is minimized. The extent of scatter is quantified by the *mean squared error (MSE),* which is mean of the squared distances on the Y-axis between a given data point and the line, for all points. (In Fig. 4.1, the distances are shown by the short perpendicular lines between some points and the best-fit line. The square is used so that it does not matter whether an individual point is above or below the line.)

The wider the scatter, the less good the fit. If very wide, you may suspect that the relationship between X and Y is not really linear at all, and wonder if another equation should be substituted. I discuss this issue next.

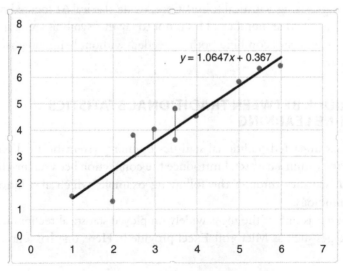

Figure 4.1 *Two-variable regression, with a straight line of best fit.* The line is chosen so that the sum of the squared lengths of the short perpendicular lines from each point to the best-fit line are minimized.

4.2.1 Polynomial regression: the overfitting problem

The same software used for multivariable linear regression can also be employed for *polynomial regression*, where the equation uses a single independent variable, X, but where there can be various terms employing powers (squares, cubes, etc.) of X. Here, in the two-variable case, the equation is of the form:

$$Y = \alpha + \beta_1 X + \beta_2 X^2 + \beta_3 X^3 \ldots \tag{2}$$

(The principle is simple: you create new variables $X_2 = X^2$, $X_3 = X^3$ and so on, so that the equation now becomes a linear regression.) It can be shown that if you have any arbitrary curve that connects N data points, you can find a polynomial with terms up to X^{N-1} and create a highly squiggly curve whose equation will fit the data *perfectly*, without any scatter at all. Fig. 4.2 shows the same data as Fig. 4.1 fitted to a sixth-degree polynomial using Excel: the scale is different because, between points 1 and 2, the curve dips way below the X-axis.

The question is whether the curve of Fig. 4.2 reflects reality. If you had a data set with a hundred data points, does it make sense to have 99 terms, plus the constant α in the equation? Further, when we apply the equation to new data points, the chances that these would lay exactly on the curve are vanishingly small.

In other words, the meandering curve of Fig. 4.2 represents a case of *overfitting*: predictions for new, previously unseen cases are poor. Similar cases of overfitting can occur if we toss in too many variables into a prediction effort. (Many analytical exercises

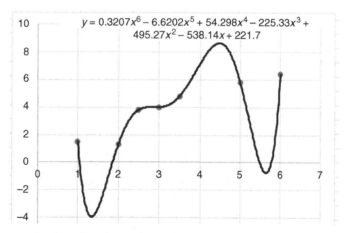

Figure 4.2 *An example of overfitting.* In a misguided attempt to eliminate the spread of data points around the best-fit line, a curve (a sixth-degree polynomial) is fit to all of the seven data points. While the fit achieved is "perfect," a new data point within the range of points is unlikely to fit on the curve. Also, between points 1 and 2, and 6 and 7, the curve goes into negative territory, even though we have no data point in our set with negative Y values.

in market research or epidemiology include every possible variable including the kitchen sink. Most of these variables turn out to be of marginal or no significance.)

With overfitting, the equation *fits the noise (ie, random variation) in the data rather than only its essential, desired features.* We have to strike a compromise between too few terms in the equation and too many. Cross-validation, which is now discussed, is a way of finding this compromise.

4.2.2 Cross-validation

One of the concerns with traditional statistics was a tendency to use *all* of the data available to fit the equation, even if an analyst did not fall into the "perfect fit" trap. Consequently, there was no systematic way to detect overfitting (or underfitting, for that matter): judgment on this issue required a mix of problem-specific knowledge and intuition. It was eventually realized that, to prevent overfitting, one must *hold some of the original data back.* The *remaining* set of data is used for determining the equation, and then the withheld data is now used to measure the equation's predictive power.

In machine learning, the set of data used for the fitting is called the *training set*, and the withheld data is called the *test set*: the test set is also called the *validation set* or *hold-out set*.

One of the standard techniques employed is *10-fold cross-validation*. One divides the data into ten equal parts. One part is withheld, the other nine parts used for the training data, and the equation coefficients obtained. Then the withheld part is used for testing, and the MSE with the test data measured. Then the second 10th-part is withheld, the other nine parts used for training, and so on, with the entire procedure repeated ten times. The test-data MSE over all ten analyses is now averaged. The experiment can be repeated with different parameters (eg, fewer variables, or more). The set of parameters with the least overall test MSE reflects the optimum solution.

Such experimental designs were tedious to perform in the hand-calculator era, but became eminently feasible as computational power increased. Also, the training/validation approach can be applied irrespective of the actual technique being employed, or the measure being employed to quantify goodness of fit.

> **The "equation" (or more typically a set of equations) may not even be directly expressed, and the user of the technique may not even be aware of them. Certain techniques, such as artificial neural networks (ANNs) or support vector machines (SVMs), which are discussed later, are somewhat like "black-boxes," in the sense that they don't readily tell you *what* they are doing: they only tell you what they "think" the answer is, and you have to quantify their accuracy based on test data to which the correct answers are already known.**

The introduction of the concept of test data versus training data marked the first advance that distinguished machine learning from traditional statistics. This idea, however, is so fundamentally important that it has percolated into modern statistics as well (though most computer statistics packages have yet to employ it routinely). Further, most of the contributions to machine learning theory today are made by mathematical statisticians rather than pure computer scientists.

4.2.3 Bias

While cross-validation is a very good way to make an ML technique robust despite variation *within* the data you are using, it does not defend against a systematic bias in the choice of data that is used for the training. If a source of data has a hidden, built-in bias, the predictions of a system that has used it for training will be unreliable when applied to the population at large. This dictum is not just true of machine learning, but of any statistical technique.

> **One of the most famous errors in the history of politics is the poll by Gallup in 1948, which indicated that Thomas Dewey, the Republican candidate, would defeat the incumbent, Harry Truman— who actually won convincingly. The source of the error? Gallup conducted the poll by telephone: at that time, only the well-off, who were more likely to vote Republican, owned telephones.**

For the same reason, a system built with data from one locale may badly misfire when applied to a different country. I remember playing with a medical expert system, ILIAD, developed at the LDS Hospital in Salt Lake City, UT, which used Bayesian calculation of probabilities (discussed later in this chapter). When supplying ILIAD with findings typical (in India) of tuberculosis, ILIAD insisted that the diagnosis was one of two fungal infections known to involve the lungs (which are indeed far more common in Utah).

4.3 A BASIC GLOSSARY OF MACHINE LEARNING

The parameters of most machine-learning problems, unlike the coefficients of linear regression, are too complex to be solved exactly. Instead, they must be solved by *iterative methods*, whose roots go back at least to the time of Isaac Newton in the 17th century. Iterative methods rely on making an initial guess as to the solution—or even setting all parameters to zero or random numbers. One then refines the guess repeatedly by applying the guess to the equation, and correcting the guess, based upon either reducing the error (such as the mean-squared error, for regression) or upon maximizing the "likelihood" (the probability of obtaining the current data, based on the current guess of the problem parameters' values).

One keeps repeating this process until the error cannot be reduced (or the likelihood maximized) any further. For complex equations with a large number of variables, the repetitions may number in the hundreds or thousands. Consequently, for realistic problems, this approach really took off only after the invention of computers, to which the drudgery can be delegated, and which can perform these repetitive calculations millions of times faster than humans (and without making arithmetic errors if designed correctly, because they don't fatigue).

The process of parameter estimation with training data is also called "*learning*"—hence "machine learning." (Ideally, learning should also include the validation steps—without these, the program has not really "learned" reliably.)

The response variable/s may be of different data types.

- *Continuous*: that is, numerical values that can be represented to arbitrary precision, but normally rounded to a specific number of decimal places or whole numbers. Examples of continuous variables are height or systolic blood pressure.
- *Discrete* data: these cannot take fractional values such as the number of children in a family. Some discrete data may fall into *categories*—such as parts of speech (noun, adjective, verb, pronoun, etc.). Other discrete variables may be *ranked*, where the categories may be graded by magnitude or severity, such as stage of cancer. Categorical and ranked data are also called *nominal* and *ordinal* data respectively. A special kind of categorical data is *binary* data, which can take only two values, such as survival/death.

Methods that try to predict the values of continuous output variables are called *regression methods*, while those that try to predict categorical values (or "*labels*") are called *classification methods*. (However, as an exception, the most widely employed classification method, which can work with continuous or discrete input variables or both, is called "logistic regression" because the equation to be solved resembles that of linear regression, as discussed later.)

4.3.1 Sensitivity, specificity, and F1

One of the advances in machine learning after it departed from the pure AI realm was more rigorous evaluation. Because of the presence of random variation in the input variables, no method of prediction can be 100% accurate. Consider the 2×2 matrix in Fig. 4.3, which should be familiar to those who have taken a course in basic clinical epidemiology. A particular method (which could be a computational, laboratory, or imaging technique) is used to predict whether a disease condition exists or does not

	Condition exists	Condition does not exist	
Prediction =Yes	Number of true positives	Number of false positives	Total (test positive)
Prediction = No	Number of false negatives	Number of true negatives	Total (test negative)
	Total (condition exists)	Total (condition does not exist)	

Figure 4.3 *An illustration of sensitivity, specificity, and positive predictive value.* These numbers apply, for example, when using a test (computer-, laboratory-, or imaging-based) to predict the presence or absence of a disease.

exist in a patient. If the prediction method is perfect, the numbers in the unshaded areas will be zero.

Some terms are defined as follows:

Sensitivity (also called *recall*) = number of true positives/total (condition exists)

Specificity = number of true negatives/total (condition does not exist)

Positive predictive value (PPV, also called *precision*) = number of true positives/total (test positive)

All of these are fractional numbers between zero and 1, though they can also be expressed as percentages.

4.3.1.1 Significance of positive predictive value

Both specificity and PPV are influenced by false positives. PPV, however, also depends on the *prevalence* of the condition in the data sample being used for evaluation of the method. Thus, if *everybody* in the sample of patients being evaluated had the disease, the number of false positives would be zero no matter how good or bad the method was, and the PPV would be 100%.

Despite this limitation, however, in many evaluations PPV (also known by its alternative name, *precision*) is employed instead of specificity. This is because the number of true negatives, on which specificity depends, is often impossible to measure accurately because the sample size is just too large. This situation arises in evaluation of Information Retrieval (Google-like) methods. If you are searching a huge collection of documents using search software while employing some keywords, it is impossible to precisely quantify the true negatives (documents not retrieved AND not relevant to the search): this would involve evaluating every "missed" document in the database (ie, the vast majority of documents) for relevance.

Recall, however, could still be estimated by identifying, in advance, a preselected and modest-sized set of documents as relevant to a given search, and seeing what proportion of these *do not* get retrieved by the keyword search: this gives us an estimate of the false-negative rate. In the lab-test scenario, one would add samples of known diseased patients to the evaluation set and see how many of these the test misses.

4.3.1.2 Determining the accuracy of a prediction method or test

By altering the prediction threshold for "Yes" or "No," it is possible to cheat and achieve 100% sensitivity, even for a completely worthless prediction method, by always reporting "Yes," even if nothing is wrong with the patient. Similarly, you can achieve 100% specificity by always insisting that nothing is wrong.

Another way of cheating is to simply use the "*majority classifier*," based on the prevalence. Thus, if the condition was very rare (eg, occurred in less than 1% of patients), the majority classifier would be "No disease," and even a worthless method that simply chose "No" every time would be correct in 99% of the population. For rare conditions like this, a new test would have to be something like 99.99% accurate to be considered worth performing. Similarly, at Mt. Waiale, Kauai Island, Hawaii, where it rains about 350 days each year, the majority classifier for weather prediction would be "rain today."

It is obvious that sensitivity and PPV/specificity tend to be inversely associated, and a true figure of merit combines them in some way. The *F1 measure* is used for this purpose. It is the harmonic mean of sensitivity and PPV, or (2 × Sensitivity × PPV)/(Sensitivity + PPV). For a method to be considered useful, the F1 measure obtained with it must be better than what could be obtained using a simple majority classifier.

4.3.2 Supervised versus unsupervised methods

Machine learning techniques can be categorized into two groups: *supervised* learning and *unsupervised* learning. In supervised learning, the values of the output variable/s are known in advance during the training phase (as in regression). The vast majority of machine learning techniques are supervised.

In *unsupervised methods*, we do not know the values of the output variable/s (or the problem may not have output variables as such), but the software organizes the input in some way.

One unsupervised approach is *clustering*, where the data is segregated into groups or clusters, such that the similarity between data points belonging to the same cluster is much higher than the similarity between points belonging to different clusters. It is up to the user to analyze the clusters in terms of domain knowledge to determine whether these clusters mean anything; thus, a population of online shoppers may be clustered based on their income level, annual dollar value of purchases, categories of items purchased, and so on.

An alternative unsupervised method is *principal components analysis (PCA)*, which was invented by Charles Spearman and refined by Raymond Cattell in the context of psychometrics (see chapter: Supporting Clinical Research Computing: Technological and Nontechnological Considerations, Section 2.2). It takes data based on a large number of input variables and tries to reduce the data into a smaller number of variables that are composites of the input variables, but which account for most of the variation in the data. Using this approach, intelligence (and personality) are categorized into multiple separate dimensions. Thus, we have verbal skills, mathematical skills, spatial reasoning, emotional intelligence, and so on. PCA is thus a form of *dimension reduction*.

4.3.2.1 Semi-supervised methods

"Semi-supervised" methods are a subcategory of supervised methods. Here, we have output values for some but not all of the data—typically because determining the output value may be very expensive or not always feasible. Data points without *Y* values are called *unlabeled*. Semi-supervised methods assume that the input data can be clustered (as for clustering), or that a set of fewer composite dimensions will account for most of the variation (as in PCA). One tries to apply one or both methods and sees whether the data points with known output values segregate cleanly. *If* they do, then one can assume that the unlabeled data points that accompany each labeled cluster most likely segregate the same way. An example is provided next.

> **In an epidemiology context, for example, suppose we have partial survival/death data on patients with a particular cancer. Suppose, on clustering the data, we find that almost all the deceased patients have a particular cell type and a particular degree of spread. Even without knowing the outcome of the unlabeled patients, we can surmise that patients with the same cell type and spread have a poor prognosis.**
>
> **Much of *human* learning is semi-supervised: we learn through a small amount of labeled experience (parental or pedagogic guidance) and a much larger amount of unlabeled experience.**

I now discuss the principles of individual methods briefly.

4.4 REGRESSION-BASED METHODS

While linear regression has been around since its invention by Pearson in 1896, the last few decades have seen a resurgence in new techniques that combine regression with machine-learning methods. The traditional regression calculation (which does *not* need machine learning to calculate) is inapplicable to the case where the number of input variables (the X's of Eq. 1) is large enough to exceed the number of data points. (In this case, you end up with a "perfect" spurious overfit to the data once the number of variables equals or exceeds the number of data points.)

Huge numbers of X's occur in many data-mining efforts today. James et al. [1] provide the example where the outcome variable, blood pressure as a function of age, gender, body-mass index, and *half a million* single nucleotide polymorphisms obtained by high-throughput screening, but with only 200 patients. Of course, the vast majority of these variables are unlikely to have any influence at all on the output variable Y. That is, their true beta coefficients are likely to be zero.

Techniques such as ridge regression and lasso regression are useful here. I am only providing the barest description here: see the text by James et al. [1] for details. These methods try to minimize, not just the MSE, but also a "penalty" that is a composite of all the individual betas of Eq. 1, multiplied by an integer, denoted "lambda," which can be varied. The two methods use slightly different composites: ridge uses the sum of the squared betas, lasso the sum of the absolute values of the betas. If some of the betas reduce to zero or become vanishingly small during the "training," the penalty will obviously reduce.

The penalty and the MSE exhibit opposite tendencies; as you increase the number of variables, MSE tends to reduce (see the overfitting example of Section 4.2.1), but the ridge or lasso penalty will increase: the optimum solution is when the sum of MSE and penalty are minimized.

Both these methods require that the individual X values be *standardized*. That is, one first finds the mean and standard deviation for each X, and then transforms each X value by subtracting from the corresponding mean and dividing the result by the corresponding standard deviation. In other words, every value is transformed into "standard deviation" units so that we have positive and negative numbers that typically range within +4 and −4. This way, we eliminate variation between the betas due to different measurement units. (Standardization is something that all statistics packages perform on demand: it is widely employed in a variety of statistical methods.)

Both methods require to be run repeatedly, gradually increasing the value of lambda from 1 upward, using cross-validation, until the fit to the data cannot be improved any further. (As lambda increases, the fit initially improves, and eventually begins to worsen. The lambda that yields the best fit is selected.)

One should normally employ *both* ridge and lasso regression. If a relatively small proportion of Xs have the major influence on Y, lasso works better. If there are many Xs with roughly equal influence on Y, ridge works better. (Of course, you don't know which of these conditions is true in advance.)

4.4.1 Using regression methods for categorical input variables

To employ linear regression incorporating variables like Race, Gender, or Disease Severity, there is a very simple technique. If a variable has N possible categories, we employ $N-1$ variables that can take either the value 0 (No) or 1 (Yes). Thus, suppose that we want to encode Race, which could take the six possible values: White, Black Native American, Asian, Hawaiian/Pacific Islander, Other. We create five separate 1/0 variables: White (Yes/No); Black (Yes/No); and so on. This allows mixed-race people to choose more than one category, and the category "Other" would be represented by the value 0/No for all five variables.

4.5 REGRESSION-TYPE METHODS FOR CATEGORICAL OUTCOME VARIABLES

Here the response variable Y is categorical, taking two or more possible values. In the simplest case, Y is binary—survival/death, success/failure, etc. In principle, one can apply linear regression, treating one value of Y (eg, survival) as 1 and the other (eg, death) as zero. (It doesn't matter whether we reverse the assignments: the results are equivalent.)

To classify a new data point, we substitute the estimated betas into Eq. 1. If the result (which is a crude measure of probability) is less than 0.5, we classify the outcome as zero, and if bigger than 0.5, we classify it as 1. This technique, called *linear discriminant analysis*, was devised (with a slightly modified formulation) by R.A. Fisher in 1936. For a long time, this was the only method used.

The problem is that sometimes the value of Y can become negative, or greater than 1. Since probability can never be negative or greater than 1, the question is how to interpret such values. Also, given a value of Y, we do not have an estimate of the probability of belonging to each class. In other words, how sure are we of our answer?

If the estimated value of Y for a new data point is much closer to 1 or 0, we are obviously more certain than if the estimate were 0.4 or 0.6, but the number that we get doesn't exactly correspond to a probability value. In cases where we are trying to predict mortality versus survival for a serious disease given a bunch of predictors, reporting a probability of survival (as well as confidence limits on our estimate) is much more useful in advising a patient (or the patient's relatives) than simply stating that the patient is more likely to live or die.

4.5.1 Logistic regression

Logistic regression, invented by the British statistician David Cox in 1958, deals with the previously mentioned problem by using a function called the *logistic function,* whose result *always* falls between 0 and 1 no matter how large (positive or negative) the result gets. The logistic function of any number, Y, is defined as $1/(e^{-Y} + 1)$, where e is the base of natural logarithms (2.78128 ...). The graph of the function against Y is S-shaped, and its value *always* falls between 0 and 1.

- If Y is a large negative number, e^{-Y} is very large, and so the value of the function approaches zero.
- If Y is a large positive number, e^{-Y} approaches zero, the numerator and denominator are, for practical purposes, equal, and so the function's value equals 1.

The logistic function turns out to be mathematically related to a function that originated in betting, called the *odds.* If an event has a probability p of happening, the odds are given by $p/(1-p)$. So if you had a 1/10th chance of winning a bet, the odds would be $0.1/0.9 = 1/9$. (In betting parlance, numerator and denominator are typically inverted, so the odds are phrased as "9 to 1 odds.") The logarithm (to base e) of the odds, also called the *logit,* happens to be the *inverse* of the logistic function. (Proving this requires high school algebra, but you can take this assertion on faith.)

In logistic regression, whose equation is related to linear regression (Eq. 1), we try to find the values of the coefficients so that the fit of the data to the equation is maximized.

$$\text{Logit}(Y) = \alpha + \beta_1 X_1 + \beta_2 X_2 + \beta_3 X_3 \dots \tag{3}$$

This looks almost exactly like Eq. 1, except that the left hand side replaces Y with logit(Y). Logistic regression can be generalized to categorical variables that fall into more than two possible classes. (Programs like MiniTab will perform multinomial logistic regression as well as ordinal logistic regression, where the categories are ranked.)

4.5.2 Support vector machines

SVMs are also used for classification, as an alternative to logistic regression (but closely related to regression mathematically). They were devised by the Soviet statisticians Vladimir Vapnik and Alexey Chervonenkis in 1963: shortly after immigrating to the United States in 1990, Vapnik and his collaborators at Bell Labs suggested the "kernel trick," discussed later.

The idea is that if the data points were plotted in N-dimensional space, a hyperplane may be identified that separates points belonging to one class from points belonging to another. In the simplest case, where $N = 2$, the hyperplane is a straight line, as shown in Fig. 4.4. Since multiple such lines are usually possible, the line is chosen that the *margin* between the line and the points on either side of it (ie, the average perpendicular distance, as shown in Fig. 4.4) is maximized.

In Fig. 4.4, the two classes (indicated by circles and diamonds) are perfectly separated by the line, but often, a straight line can only be drawn if a small minority of items fall on the "wrong" side of the line (ie, get misclassified.) In this case, you have to strike a compromise between maximizing the margin and the proportion of misclassified items.

The advantage of SVM over simple linear regression using a 1/0 Y variable is that linear regression uses all the data points in the calculation of the line of best fit, while SVM chooses to focus on the set of points closest to the margin (this set of points is called the "support vector"). In Fig. 4.4, this set is indicated by the dark-shaded circles and diamonds.

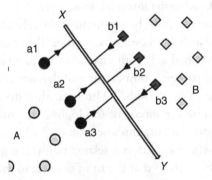

Figure 4.4 *An illustration of an SVM in two dimensions.* Points belonging to two different classes (indicated by *circles* and *diamonds* respectively) are separated by a line, which is chosen so that the "margin" between the classes is maximized. The two classes could represent outcomes for a disease (eg, cure versus no cure) and we would like to predict outcome based upon multiple variables. Unlike regression, which uses *all* the data points, an SVM only uses the data points *closest* to the margin (indicated by the *dark circles* or *diamonds*). The points closest to the margin are called the "support vectors."

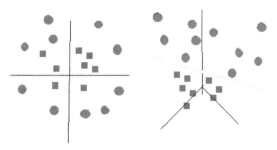

Figure 4.5 *An example of the "kernel trick" to allow linear separation between classes.* As in Fig. 4.4, there are again two classes of points, indicated by *circles* and *squares*: however, as shown on the left, the squares are entirely surrounded by the circles in two dimensions, and no straight line can be drawn to separate the two. On the right, we compute an additional variable for each point—the radial distance from the origin, and plot each point in *three* dimensions, with the radial distance plotted on the Z-axis. The circles now "float" above the squares, and a plane can be constructed (the *thick pale line* shown edge-on) that separates the two points. The radial distance is an example of a "kernel function." (A real-life counterpart of this phenomenon is where extreme values of an electrolyte level—for example, either extremely low *or* extremely high serum potassium levels—are associated with poorer outcomes than normal or mildly altered levels.)

Sometimes, however, the two classes cannot be directly separated by a straight line, as shown on the left-hand part of Fig. 4.5, where the squares lie entirely inside the area of the ellipses. In such a case, one introduces a new variable that is a transform of the data, and plots the data incorporating a higher dimension. In the present case, the radial distance from the origin can be employed. (In two dimensions, the radial distance of a point with coordinates (x, y) is $\sqrt{(x^2 + y^2)}$.) The points are plotted in 3-D instead of the original 2-D, and the radial distance is plotted on the Z-axis. Now, the ellipses appear to "float" above the squares, since each ellipse's radial distance is greater than each square's radial distance. A separating hyperplane (indicated by the pale broad separating line, which has been depicted edge-on) can now be constructed.

The process of incorporating a mathematical transform, called a "kernel function" (and a higher dimension to incorporate its value) is called the "kernel trick." (One does not actually have to work in the higher dimensions: so using a kernel function, while taking extra computation time, does not expand memory requirements greatly.)

Technically, a kernel is a mathematical function with two properties:
1. **Its shape is symmetrical about the *Y*-axis—that is, its left and right halves are mirror images.**
2. **The area under the curve is exactly 1.**

There are several functions that can be used as kernel functions. The choice depends upon the problem: SVM software will let you try out different functions and see which one gives the best separation.

4.5.3 SVMs versus logistic regression

Like logistic regression, SVMs can also be generalized to categorical output variables that take more than two values. On the other hand, the kernel trick can also be employed for logistic regression (this is called "kernel logistic regression"). While logistic regression, like linear regression, also makes use of all data points, points far away from the margin have much less influence because of the logit transform, and so, even though the math is different, they often end up giving results similar to SVMs.

As to the choice of SVMs versus logistic regression, it often makes sense to try both. SVMs sometimes give a better fit and are computationally more efficient—logistic regression uses all data points but then the values away from the margin are discounted, while SVM uses only the support-vector data points to begin with. However, SVM is a bit of a "black box" in terms of interpretability. In logistic regression, on the other hand, the contribution of individual variables to the final fit can be better understood, and in back-fitting of the data, the outputs can be directly interpreted as probabilities.

4.6 ARTIFICIAL NEURAL NETWORKS

ANNs got their name because they began as a simulation of how brain cells were believed (in the 1950s) to work. ANNs comprise one or more "layers" of autonomous computational units that received input from other units (including units in the same layer), and send outputs to still other units (or even feedback to the cells that provide input). The earliest ANNs, called *perceptrons*, were used for classification tasks, just like basic SVMs or linear discriminant analysis. Eventually, it was realized that you can stretch the analogy only so far, and ANNs should be regarded as a form of computation that converges iteratively on a solution for a classification problem: they continue to be used in production for handwriting recognition, especially zip codes on US mail— where they enable automatic sorting.

Training (ie, supervised learning) and cross-validation steps are employed with ANNs as with any other machine-learning method, and during the training the internal parameters that characterize individual computational units are gradually tweaked—and this is done automatically by the software, as for most machine-learning methods—until an optimum is reached, in terms of sufficient sensitivity/specificity for the classification task. ANNs can also be used in unsupervised mode, to do automatic clustering and principal components analysis of data.

At one time, ANNs were much more popular than they are today. The following disadvantages are now known:
- When applied to the same category of problems that linear regression and SVM (which came much later) are suitable for, the latter two run much faster.
- Linear regression also provides more explanatory power, in terms of identifying which input variables are more important than others—and which can be discarded because they don't contribute to the classification. With ANNs, it tends to be a case of "the number I'm scanning is

a six because I say so." The internal parameters are opaque to the typical user. Also, because you end up using many more input variables than may be truly necessary, and not know that you are doing so, there is a much greater risk of overfitting compared to the regression methods.

- "Vanilla" ANNs—where the network is simulated entirely in software—are extremely compute expensive compared to the alternatives.
- Sometimes, ANNs converge prematurely on a solution that is not the most optimal one. (By analogy with exploring unknown undersea terrain, where you are searching for the deepest point, ANNs will tend to converge on a local hollow rather than the very deep trench a mile away.) In technical parlance, ANNs tend to converge on the "local minimum" rather than the "global" minimum.

Some of these problems have been circumvented in several ways.

- ANNs are implemented in custom hardware, which is specialized for a particular task (such as mail sorting). This approach makes sense only when the hardware is going to be employed industrially on a large scale.
- ANN software is now being optimized by taking advantage of microcomputer graphical processing units (GPUs), which are also employed by video game software (especially that which involves lots of rapid action). Of course, alternative ML software is also being rewritten to leverage GPUs.
- Escaping local minima is done by employing techniques such as "simulated annealing" or "genetic algorithms." Using these approaches, instead of the iterative training process invariably seeking a *better* solution than the previous one, a certain proportion of the time, the software will select a candidate solution that is *worse*. This is the analog of moving away from a local hollow with the idea of more comprehensively exploring the entire space. Obviously, computation time greatly increases, but the final solution may be much better. (Microsoft Excel 2010 and later gives you the option of employing a genetic algorithm for the Excel Equation Solver.) These methods of escaping local minima are not unique to ANNs—they can be used with almost any ML method. See the Wikipedia articles on these topics if you want more information.

ANNs are best reserved as a backup strategy for problems where logistic regression or SVMs, after being tried first, don't give good results. An advantage here is that ANNs are completely agnostic about the equation or set of equations that fit the data—unlike, say, linear or logistic regression, which assume specific equations that may not apply. While taking a long time to converge on a solution during training, ANNs will ultimately do so.

Because ANNs have been around for a long time, commercial ANN software has been commoditized to be very user-friendly.

4.7 BAYES' THEOREM AND NAÏVE BAYES METHODS

Bayes' theorem, named after the clergyman/mathematician Thomas Bayes (1710–1761) and published after his death, is used to calculate the probability of something happening, based on conditions that might be related to that event. The clearest expositions of this all-important theorem that I've seen are in Provost and Fawcett [2] and in Nate Silver's classic, *The Signal and the Noise* [3].

In the next equation, the term H indicates a *hypothesis* (eg, that a patient has a particular disease), and E indicates *evidence* related to that disease (eg, patient has a rash). There may be other causes of the rash besides the specific disease, and all patients with this disease may not have the rash. The terms $p(H)$ and $p(E)$ indicate the probability of the hypothesis, and probability of evidence, in the population respectively. (In the present case, these probabilities can be calculated by simple tallying if we have data on the prevalence of the disease, and data on the prevalence of the rash). The term $p(E|H)$ is the probability of the evidence given the hypothesis. (This could also be calculated directly, for all patients with the disease diagnosed by an alternative, highly accurate method, if we tally how many of them have a rash.)

Bayes' theorem allows us to calculate the probability of the hypothesis given a case where the evidence is present. It is formulated as

$$p(H \mid E) = p(E \mid H) * p(H)/p(E) \tag{4}$$

In the present case, $p(H|E)$ is the probability that, if a patient shows up with the rash, then he or she has the specific disease.

In the real world, many different kinds of evidence may be brought to bear on the same hypothesis. In the previous example, alternative evidence for the same disease may include the presence of diarrhea, the presence of cough, and so on.

4.7.1 Conditional independence and Naïve Bayes

Two sources of evidence are *independent* if neither one influences the other. (Thus, the food preferences of different, unrelated individuals living far apart, in normal circumstances, do not influence each other.) Two or more evidence sources are assumed to be *conditionally independent* when they bear upon a *common hypothesis*, but knowing about one evidence source does not give us any more information about the other. As an example, given the presence of disease, *if* the evidence of diarrhea in a given patient does not make the presence of rash, more or less likely, and vice versa, then we say that, for this disease, the findings of rash and diarrhea are conditionally independent of each other.

Very often, the assumption of conditional independence is tenuous—you don't know for certain that two sources of evidence are conditionally independent until you verify that with a large amount of data (and maybe prospective experiments). However, this assumption can make calculations dramatically simpler by ignoring the "interactions" between evidence sources, assuming that they are too small to matter. The assumption of conditional independence, even though such an assumption has not been verified, is called *the Naïve Bayes assumption*.

For example, what is the probability that a patient has the disease, given the presence of rash AND cough AND diarrhea? Using Naïve Bayes, one employs the simplified formula

$$P(\text{disease} \mid \text{rash, cough, diarrhea}) = P(\text{rash} \mid \text{disease}) \times P(\text{cough} \mid \text{disease})$$
$$\times P(\text{diarrhea} \mid \text{disease})/[P(\text{rash}) \times P(\text{diarrhea}) \times P(\text{cough})] \tag{5}$$

All the numbers in the right hand side of the equation can be estimated from individual prevalence data (which really consists of simple counts): we do not have to know how many patients in the overall population have combinations of individual pairs of symptoms. This simplification can be significant in cases where there are a very large number of evidence sources that bear on the same hypothesis, as discussed shortly.

Naïve Bayes is also a classification technique like logistic regression and SVMs, described earlier. It is useful when most or all of the predictor variables are also binary or categorical (unlike SVM and logistic regression, which are preferred when the predictor variables are mostly numeric). It deliberately sacrifices some accuracy in prediction to make the computations involved tractable in the first place. For example, Naïve Bayes is the most widely employed technique for the detection of spam e-mail. Individual units of evidence may comprise presence (or absence) of the terms "Viagra," "Cialis," "Nigeria," "No down payment," "Easy instalments," "Porn," "Your account has been suspended/disabled/locked," and so on. The list of terms that may possibly indicate spam can number in the hundreds or thousands. When small numbers of probabilities get multiplied, the final numbers can get vanishingly small, and you may get a phenomenon called *underflow*—where, if two numbers are small enough, the computer reports their product as zero. In such cases, a technique commonly employed is to take the logarithm of the probability (which is a negative number), and add the logarithms up (which is the equivalent of multiplying the raw probabilities).

Naïve Bayes can also be generalized to the case where instead of only two outcomes (eg, disease present/absent, spam/no spam) we have three or more outcomes (eg, different types of leukemias). Here, the approach is to calculate the probabilities of each of the outcomes, and then choose the outcome with the highest probability. However, when the assumption of conditional independence is violated, the calculated probabilities should be considered as ballpark figures rather than very precise numbers, and so approaches based on Naïve Bayes suffice only for *ranking* answers rather than using the probabilities verbatim.

Naïve Bayes has the advantage of incremental learning—because the computations are so simple, only the new data need be brought to bear. This makes it eminently suitable for "Big Data" efforts, discussed later.

4.8 METHODS FOR SEQUENTIAL DATA

A special class of machine-learning methods is employed on *sequential data*. Here, the input consists of either a single continuous variable whose value is changing continually with time (eg, an electrocardiographic, audio or electroencephalographic signal), or a sequence of discrete features—such as words or phrases in a passage of text, nucleotides or amino-acids in a DNA or protein sequence respectively. In some cases, you want to identify the base components that account for the signal: in other words, you want to perform *feature extraction*.

Feature extraction is performed by unsupervised techniques such as *Fourier analysis* (Section 2.2), which tells you what individual frequencies exist in the underlying signal, or *wavelet transforms,* a more powerful, though less compute-efficient technique employed when the frequencies themselves change with time (an example is a siren that ramps up or dies down). Prateek Joshi's blog, "Perpetual Enigma" [4], explains both of these concepts very clearly.

Mr Joshi is a software developer/researcher who works in image processing and machine learning, among other things, and is currently based in San Jose, California. He excels in explaining difficult topics in a manner that is accessible to the nonmathematician or noncomputer scientist. On his blog, which is searchable by keyword, he quotes Albert Einstein, "If you can't explain it simply, you don't understand it well enough." I've learned a lot by reading his nutshell explanations. If you're the sort of person who gets brain freeze looking at reams of equations and mathematical symbols—many articles in Wikipedia are unfortunately guilty of this crime, providing insufficient examples or overview to the nonspecialist—you will find his blog very useful.

4.8.1 Markov chains and N-grams

Proposed by Russian mathematician Andrey Markov in the late 19th century, a Markov chain refers to a probabilistic sequence of events (also called "states") where the probability of an event is influenced by a finite number of *previous* events, each of which have their own probabilities of occurring. The number of previous events is typically modest because the computations involved can otherwise get very complex.

In the extreme case, the number is 1, which means that the *probability of a state occurring depends only on the previous state.* (Alternatively, the probability of the *next* state depends only on the *current* state.) This assumption (called the *Markovian assumption*) may not always be true. However, as long as the effect of more remote events is much less than that of the previous event, the assumption has the effect of making the calculation of the probabilities of a sequence of events considerably simpler: you just multiply the individual probabilities—just like Naïve Bayes.

When applied to natural language, the "events" are the occurrences of particular words or word categories in text, which are followed by other words, and so on. If large volumes of text are available, the probabilities can simply be "learned" from the data, though some can be inferred by rules of grammar. Thus, the word "the" is always followed by a noun phrase, and "Mr." always followed by a name.

Google stores the millions of queries that users ask and creates "N-grams"— sequences between two and five words, and the frequency with which they occur in queries. Also, as part of the Google Books project, Google has computed N-gram data from the content of tens of thousands of books: this data can be queried freely at https:// books.google.com/ngrams. You can enter phrases of between one and five words and observe how the frequency of that phrase has changed over time (1800–2000) in printed content. (Try entering "Huckleberry Finn")

The "N-gram" data and the accompanying frequency information can be used for various purposes.

1. *Autocompletion*: (also used in Wikipedia) greatly improves the user experience by anticipating what word the user might type next. (Try typing "electronic health" into Google or Wikipedia's search box: even before you finish typing "health," the word "record" will be suggested.)

2. *Spelling correction*: If a word is not recognized due to misspelling, but neighboring words are correctly spelled, the software can guess at what the correct word might be, based on the N-gram data. (Which N-grams include the previous and next word, one word apart? Which words are in the middle? Find the word that is most similar to the misspelled word.)

3. *Speech recognition/sound disambiguation*: The same strategy as earlier can be used to resolve homophones ("two" vs. "too"). "Too" is followed by an adjective, "two" by a noun.

4. *Word disambiguation in written or spoken speech*: an extension to the previous strategy. A phrase with multiple meanings (eg, "cold") can be disambiguated: cold (temperature) is followed by nouns related to weather of climate, or preceded by a linking verb (is, was). Cold (rhinitis) is associated with words related to illness, or preceded by an article. (With enough data, software could resolve "Because it was cold, I got a cold.")

5. *Parody text generators*: These computer programs use a large collection of documents (eg, computer science journals), generate N-grams from the data, and use that information to generate prose. The prose appears fully grammatical, but is actually meaningless. Essentially, you start by picking a word at random, then randomly pick another word that follows the first word in the collection frequently, then randomly pick a third word that follows the first two words in the data, and so on. The bigger the collection of documents, and the lengthier the lookback employed during text generation, the more realistic the text appears. (That is, if you use 4-grams or 5-grams, the text is more realistic than if you use only 2-grams.) Jon Bentley describes the construction of such a program in his delightful (and classic) collection of essays, "Programming Pearls" [5].

The Postmodernism Generator [6] automatically generates an essay in post-modernist prose. To make the output more realistic, these programs also use a context-free grammar (a set of rules similar to those used to implement programming languages), so that there are no obvious grammatical errors. A parody-generator of computer-science papers, SCIGen [7], created by three MIT graduate students as a practical joke, generated a paper that was accepted at a "spamference"—a supposedly scientific conference whose only purpose is to make money for its organizers, and lengthen the curriculum vitae of those who submit papers to it. (The organizers get hold of mailing lists of researchers to whom they send bulk e-mail with requests for submissions.)

Interestingly, there exist numerous predatory journals in the same racket, which will accept any submission without even glancing at its contents, as long as a "processing fee" is paid. One of these accepted, in 2014, a paper that consisted only of the sentence "Get me off your ----ing mailing list" repeated endlessly. The paper itself was a "plagiarism" of a "manuscript" created by two researchers (from Stanford and Harvard) in 2005 to respond to repeated unsolicited conference/journal invitations [8].

N-grams are relatively simple to implement. However, to be ergonomically useful, they require churning through vast amounts of data, and may be considered the result of continual data mining.

4.8.2 Hidden Markov models

For discrete features (or features that have been discretized, such as the individual sounds that comprise speech), certain supervised methods draw inferences about the underlying events or states (also discrete) that are likely to account for a sequence of observations.

- For text, the "observations" are words or phrases in text, the "states" may be parts of speech or semantic categories. Thus, "go" is a verb, and "dog" falls in the category "noun."
- In speech, the observations are sound signals. The states are the words that give rise to the signals. Thus, the same sound, "tu:" could represent any of "two," "to," or "too." Also, the same word may be pronounced differently by different people: British speakers of English and New Englanders tend to drop their trailing "R"s, for example. In other words, the same state could result in (slightly) different sounds.
- For a protein sequence, the observations are individual amino acids. The states are the functions of individual segments: metal-binding site, DNA-binding site, cofactor binding site, etc.
- Depending on the problem, the number of possible individual states, and the number of possible individual observations, could both range in the hundreds or thousands.

Hidden Markov models (HMMs) [as well as *conditional random fields* (CRFs), discussed in the next section], are supervised (ie, data driven) methods that try to *infer the states* (which may be treated as a chain of Y variables) from the *observations* (which are X variables). Both address the same problem, but use different approaches and make a slightly different set of assumptions.

The term "hidden" in hidden Markov refers to the fact that, though the states give rise to the observations, the states are "hidden" and have to be inferred.

Let us focus on the speech recognition problem. For this purpose, we need a large collection of spoken sound signals (also called "phones," corresponding to individual syllables) that have already been tagged with the individual words corresponding to them (by humans, of course). Individual words correspond to a particular sequence of one or more phones. The signals must originate from different speakers, so that when the system is trained, it accommodates some interspeaker variance.

In other words, our training data consist of very large numbers of sequences of X values (observations) with their corresponding known Y values (states). Then, when the system hears a sequence of phones/X values from a new speaker (*even if the exact sequence of Xs has not been encountered before in the training data*), it must try to infer the sequence of Y values (words) that most likely correspond to the sequence. If we succeed in doing this, we have achieved speech-to-text translation.

As in the case of N-grams, certain words tend to be followed by other words: that is, certain sound tend to be followed by other sounds. The probabilities of these can be computed from the training data.

4.8.2.1 Basic assumptions in employing an HMM

In the HMM, there are several basic assumptions that are used to dramatically simplify the computations involved.

1. The Markovian assumption: the probability of switching from a current state (Y value) to the next state depends only on the current state. (This may not strictly be true, but for speech, it happens to be "good enough.") Given a particular Y value, there is usually a limited choice of succeeding Y values, each with a different probability. Thus, in English (though not in Ukrainian), the T sound (without a subsequent vowel sound) is never followed by a "K" sound, and in English (though not in Sanskrit-derived languages such as Hindi), "K" without a succeeding vowel is never followed by "SH"

2. The probability of generating a particular X (eg, sound) from a particular Y (eg, intended written syllable) depends only on the corresponding value of Y, not on previous or succeeding Ys. (Again, this is usually a "good enough" assumption.) The possible Xs that could be generated from a given Y are limited. Thus, the initial R in the one-syllable "rose" may be pronounced with a trilled "R," like a "W" or like an "L" by native Scottish/Irish, French, and Chinese speakers respectively. Both of the previously mentioned sets of probabilities can be derived from training data.

   Speech recognition systems generally don't do too well with diverse accents, because there is not enough training data to suggest with a sufficiently high probability that, for example, "R" in an English recognition system could be a candidate for an "L" sound.

3. The individual states (Y values) are conditionally independent of each other. In other words, if the probability of the sequence ($Y1$, $Y2$) is A, and the probability of the sequence ($Y2$, $Y3$) is B, then the probability of the sequence ($Y1$, $Y2$, $Y3$) is $A \times B$. (In other words, we can employ a Naïve Bayes strategy to calculate probabilities.)

4. The individual observations (X values) are conditionally independent of each other.

   (In the case of speech, assumptions 3 and 4 are "good enough" rather than strictly true.) Because of the assumption of conditional independence, *the HMM is an application of Naïve Bayes to sequential data.*

4.8.2.2 Subproblems with a hidden Markov model

Given a set of S distinct states, and a set of O distinct observations, we have three sets of probabilities.

1. The set of probabilities that the first state in the Y sequence—the starting symbol—will have a particular value. (In speech, an individual sequence may be a sentence, and from the training data, one can directly calculate a histogram of all starting symbols in the data. In written English, for example, a verb is not likely to start a sentence unless it is a directive (eg, "Eat your vegetables!" or "Go home").

2. The state-transition matrix (of size $S \times S$). A given cell in the matrix indicates the probability of switching from a given state $Y1$ to another state $Y2$. (Many of the cells in the matrix may be zero: for example, in English, a syllable ending with "F" could never be followed by a syllable starting with "D.")

   Note that a given state may be followed by the *same* state, as in the English "oh oh" or its Mexican equivalent, "ay ay ay." In English speech, cases where $Y1$ is followed by another $Y1$ are uncommon, though in Malay, two-syllable words are often repeated. In other cases (eg, weather), where Markov models have also been applied, they are not so rare. A sunny day is most likely to be followed by another sunny day, and a rainy day may often be followed by another rainy day.

3. The output or "emission" matrix (of size $S \times O$). A given cell in the matrix indicates the probability of a state $Y1$ generating an observation $X1$. Again, most of the cells may be zero. (For example, a syllable beginning with a vowel could never output a sound beginning with a consonant.)

The probabilities are calculated from the tagged training data by maximizing the likelihood of the fit of the estimated probabilities to the data. Obviously, for a situation where you have thousands of states and probabilities, you need a very large amount of training data for the probability estimates to be reliable. (The algorithm that performs the maximum-likelihood fit is called the Baum–Welch algorithm.)

Once we have an estimate of the probabilities mentioned earlier, we can deploy the HMM into practice. There are two different problems that can arise.

1. Given an unknown sequence of observations, calculate the probability of a particular sequence of states that might have generated it. (For example, given an observed sequence of sounds, how likely is it that a particular word/sequence of words corresponds to it?) This can be done by using the probability matrices described earlier, using a series of multiplications.

2. An extension of the previous point: Given an unknown sequence of observations, which is the *most likely* sequence of states that might have generated it? One has to find all the candidate state sequences that could lead to the given observation sequence, and then pick the one with the highest probability. The challenge is doing this in real time, by rapidly discarding very-low-probability solutions. (Speech recognition must work as the speaker continues to talk: HMMs are also used in telecommunications—such as in your cell phone.)

The algorithm to solve problem 2 was invented by Andrea (Andrew) Viterbi, a cofounder of the telecommunications giant Qualcomm, who implemented it in circuitry, making enough money to donate $52 million to the University of Southern California, which renamed its School of Engineering after him. Viterbi had immigrated to the USA in 1939 as a 4-year-old refugee with his Italian Jewish parents.

Of course, since the speech-recognition process is probabilistic and is driven by the size and diversity of the training data, it can never be 100% accurate. If the system hears a proper name it has not encountered before, such as that of a foreign person, the word corresponding to that phone sequence will not be in its "database," and so the speech-to-text output will be mangled. But then, human listeners will also do the same thing. How many humans, hearing for the first time the sound of the last name of the famed Duke University Basketball coach, "she-shev-ski" would guess that it is spelled "Krzyzewski"? The probability of this interpretation, based on the training data (assuming it originated entirely from the USA rather than Poland) are miniscule to zero.

Note that in the HMM problems, we only have two variables—the observations X and the states Y. Both are categorical variables (with potentially thousands of categories for certain real-life problems), but there is only one set of Xs. *In other words, the observation categories are mutually exclusive: that is, a given observation cannot take two values simultaneously.*

4.8.2.3 HMM issues with multiple features

What do we do if there are multiple variables in the data? For example, in processing of text, a phrase may be a part of speech; it may/may not contain a prefix, a suffix, an embedded number, capitalization, or an embedded hyphen. All of these are separate features. Thus, the word "disassociate" is a verb, also contains a prefix (dis), has a word length of 12 letters and is not hyphenated. Similarly, in a computer vision problem (recognition of objects), we use information such as shape, color, contrast within the shape (eg, the stripes of a zebra vs. the uniform color of a donkey), and so on.

In other words, *for many real-life problems, a single observation may have multiple, separate features.* HMMs can (in theory) deal with this by creating artificial variables that are permutations of all the individual possibilities, but there are two problems in doing so.
1. The universe of categories has expanded greatly; the math can now become intractable.
2. The assumption that the observation categories are conditionally independent of each other (assumption 4 in Section 4.8.2.1) is not true. (For example, a capitalized word in the middle of a sentence is always a noun or noun phrase.) If we choose to ignore such dependencies, just so that we can do the calculations, the accuracy of the HMM's predictions suffer.

The next section describes a method that can deal with this situation robustly.

4.8.3 Conditional random fields

CRFs address the same problem as HMMs. The difference is that, as stated previously, they are suitable when an observation consists of a set of features. Using training data, CRFs calculate the probability of individual values of the categorical Y (state) variable using logistic regression. As discussed earlier, logistic regression is the standard method when the input variables are categorical or numeric, and the output variable is categorical.

There is also a slight shift in terminology. The HMM model assumes that the hidden states (Y) "generate" the observations X. CRF makes no such assumption: all that it does is assume a given sequence of Y values can be *predicted* from a given sequence of data points, *each* of which has multiple features, $X_1, X_2 \ldots Xn$, where n is the total number of features.

The following are advantages of CRFs.
1. With logistic regression, it does not matter if certain features are highly correlated with each other; the regression will still work. Further, with approaches such as stepwise regression, where we include one feature at a time and repeatedly perform regression, features that do not contribute significantly to the final fit may be dropped. (For example, in using CRFs to guess a part of speech for a phrase not previously encountered before, word length may turn out to be unimportant.)
2. Unlike HMMs, a given value of Y is permitted to be influenced by *several* consecutive values of X rather than just one. Thus, in the problem of object classification in text, "San Jose" by itself would indicate a Y value of "place," but "San Jose Mercury" would refer to a newspaper. Similarly, in word-meaning disambiguation, the disambiguating words or phrases may be far away from the ambiguous phrase.

For example, the phrase "Jack shot a buck" can either mean "Jack used a bullet to injure or kill a male deer," or "Jack lost a dollar." By itself, the meaning cannot be resolved. However, phrases in the neighboring sentences like "hunt," "rifle," "bet," "gamble," etc. would suggest one meaning or another.

Note that the phrases that enable disambiguation could occur at a considerable distance *after* the problem word. N-grams, an alternative approach to disambiguation described in Section 4.8.1, work only when the disambiguating term is close by—even Google, with its vast computational resources, limits itself to 5-grams at the most.

What about the choice between HMM and CRF where there is only one input feature? Work by Ng and Jordan [9] (comparing the underlying approaches, Naïve Bayes versus logistic regression) indicates that, as the training data gets larger, logistic regression ultimately achieves a better fit, but that Naïve Bayes "learns" from the data much faster. Another advantage is that the logistic-regression approach of CRF doesn't merely tell you which the "most likely" sequence state is: it also outputs a "confidence measure"— in other words, it tells you how sure the CRF model is that this is the correct answer. Therefore, based on the risk of misclassifying—which varies with the application—you may accept or reject its recommendations.

> Risk, which quantifies the consequence of false positives/negatives, is important in many applications. On Sep. 6, 1983, a Soviet early-warning satellite system signaled an alarm indicating that the USA had launched five intercontinental ballistic missiles. The sensors, which were aimed at the known location of a US missile field, looked at the edge of the earth, with the idea that missiles that had risen 5–10 miles up would be silhouetted (by reflecting sunlight) against the blackness of space. This might have triggered World War III. Fortunately, Soviet Lt. Colonel Stanislav Petrov, who was in charge of the listening station, had the guts to doubt the sensors and double check. He figured that if the United States was about to start a war unilaterally, they wouldn't start it with only five missiles. [10].

4.9 INTRODUCTION TO NATURAL LANGUAGE PROCESSING

Natural language processing (NLP)—the process of extracting information from ordinary prose—has its roots in computation (especially AI techniques) applied to linguistics and grammar, going back to the 1950s. Many of the tasks involved in NLP are identical to those in information retrieval, discussed in the previous chapter: dividing a document into sections, splitting on sentences, tokenizing words, identifying root forms, and flagging ambiguous terms. Both NLP and IR may employ concept recognition (if you have an electronic thesaurus to identify synonyms): this task is made more difficult by lexical variants (eg, order of words in a phrase, plurals), ambiguous words with multiple possible meaning (ie, homonyms) and abbreviations (which can also be ambiguous).

One task specific to language processing is *parsing*, which means taking text in any form and converting it into an internal computer representation that lets you do something useful with it. Parsing is the first logical step in language processing. Parsing relies on a set of *rules*, which collectively comprise a *grammar*. For example, rules of grammar for a language like English specify how a unit of text (typically a sentence) must be decomposed into parts of speech, subject, object, etc. An ungrammatical construct is one that violates one or more rules of the specified grammar.

The idea of getting humans to communicate reliably and productively with computers (as opposed to flipping electronic switches, or later, writing instructions specific to a particular machine) took off with the invention of the first "high-level" programming language, FORTRAN, which was an acronym for FORmula TRANslation. FORTRAN was designed at IBM by a team led by John Backus. The FORTRAN parser was initially designed in a completely *ad hoc*, "seat of the pants" fashion, without any underlying theory to guide the process. In 1956, Noam Chomsky, who is now a well-known social critic, devised a classification of grammars, from most restrictive to least restrictive—in terms of what the user of the language employing the grammar is permitted to do. This work has had widespread impact.

- At one extreme (most restrictive), you have "regular" grammars, used to specify text search patterns with punctuations and symbols. In PubMed, for example, the wildcard symbol (*) indicates a root phrase—thus, *comput** would pick up *computer/s*, *computing*, and *computational* while the symbols AND/OR indicate Boolean search. Regular grammars are the easiest to implement. A tool called *lex* (for lexical analyzer generator), which generates code automatically from a regular grammar specification, was invented in 1975. Enhanced versions of it continue to be used.
- At the other extreme, you have "context-sensitive" grammars like English and other natural languages, where what is considered grammatical (or not) varies with context. This arises because a single word can have multiple meanings and even can represents different parts of speech at different times: "leaves" can be a noun (plural of "leaf") or a verb (present tense of "gone away"), while "can" could mean "is able to" or "a metal container for packaging food or drink." Context-sensitive grammars are extremely challenging for computers to handle.
- Grammars intended for programming languages are intermediate in complexity and are called "context-free," where the rules never change, and words specific to the language can mean only one thing. These can be described by Backus–Naur form (BNF), a means of formally specifying a context-free grammar as a set of rules, devised by John Backus (given earlier) and Peter Naur. Software tools (such as *yacc*, also created in 1975, and used along with *lex*) will convert a set of rules expressed in BNF (or something very similar to it) into a programming language interpreter or compiler (both of which convert code written by a human into instructions that can be executed by a machine).

4.9.1 Eclipse of manually created rule-based approaches

Ultimately, the purpose of NLP is to extract meaning from text. While implementation of programming languages relies critically on rule-based approaches, where the grammar is hand crafted by a human expert, they have proved unwieldy for natural language

- They are labor intensive. More and more rules have to be create to catch violations of "common sense"—such as the fact that you can play a game (or music), but not a book. (Chomsky coined the phrase "Colorless green ideas sleep furiously" as an example of grammatically correct but semantically nonsensical sentence.)
- Many of the rules of grammar are not essential to understanding spoken or written communication. Natural language has a lot of *redundancy*, that is, we use many more words than we have to. Telegrams and telegraphic clinical notes can be understood even without using articles ("the," "an"). Also, poetry can usually be understood even if it violates most grammatical rules. Systems that rigidly rely on grammatical structure to extract meaning tend to break down in these situations.
- They are brittle. It is hard to implement purely rule-based systems that degrade gracefully when the input text's complexity exceeds the system's knowledge.
- Ambiguous phrases make extraction of meaning harder unless you consider *probabilities*, which the purely symbolic approaches do not. For example, when encountering ambiguous phrases, CRFs, discussed in the previous section, use the presence of specific words to gradually increase the probability of one interpretation over another. In the absence of more information, one interpretation may be favored over another based on statistical frequency in a large volume of annotated data.

We find puns funny because they deliberately make use of ambiguous parses, where the trigger phrase suggests more than one interpretation (eg, when vocalized), or when the interpretation suddenly switches to the less probable one, as in Groucho Marx's, "I shot an elephant in my pajamas. How he got in my pajamas I'll never know." Purely symbolic parsers, however, might throw a fit.

For a long time, NLP stayed away from numbers of any kind. It was realized eventually that to make progress, you had to consider probabilities. (This is what humans do in a crude way, considering the most likely of several interpretations when encountering ambiguity.) This realization led to the birth of statistical NLP, which arose in the 1980s. Statistical NLP exemplifies Rajaraman's dictum (discussed in the chapter on data mining): more data usually beats more elaborate algorithms.

For example, in statistical parsing, rather than trying to devise thousands of rules for every special case manually, one uses much fewer rules associated with probabilities, which are learned with machine-learning programs (such as those based on HMMs and CRFs) with large volumes of annotated data. (In other words, statistical NLP mostly uses supervised learning.) This data then allows the software to come up with the most likely interpretation of individual sentences (based on the data). Alternatively, it will suggest multiple options, with the probabilities of each. (In fact, the Viterbi algorithm, mentioned in the HMM section, is employed here to find the most probable parse.)

Manual rules are still used, but judiciously so: when combined with statistical methods, results seem to be better than with either alone, at least in medical NLP [11]. Additionally, some features/input variables for machine-learning approaches are based on manual rules.

While annotation of the data for statistical NLP must be done manually, of course, the more data you have, the better the statistical approach gets. The major caveat here is that the data must be representative of the problem to be solved. In other words, you can't use newspaper articles as raw material to train a program that must interpret medical text. On the other hand, it is not advisable for your data to be too narrow (eg, having documents related only to a specific disease): otherwise, as in machine-learning methods in general, your solution, while giving astonishingly good results for that disease, will be overfitted and will not generalize to another disease. Unfortunately, many researchers getting into the field and hell bent on publication do just this, devising an NLP approach (using machine learning or hand-crafted rules) that works only for, say, chronic leukemias, simply because the precision/recall/F1 numbers they get look so good.

A good (if quite technical) book on statistical NLP is Manning and Scheutze [12]. This text, however, is long overdue for a revised edition.

4.9.2 Information retrieval versus NLP

In the past, IR and NLP were like oil and water, and the fact that there are still separate textbooks on each topic indicates that they haven't been fully synthesized. However, the differences between the two are gradually being blurred. Limited use of NLP techniques can improve the accuracy of IR for medical text in specific circumstances, allowing the following.

- Recognition of negation (the concept is absent or checked for and ruled out): for example, the vast majority of occurrences of the word "fracture" in clinical notes are cases of suspected rather than confirmed fracture.
- Gradation of severity, in cases where the parameter being analyzed can be graded (eg, pain, obesity), or subcategorization (eg, smoking status can be one of nonsmoker, ex-smoker, or current smoker).

There are several practical problems with NLP, however.

- It is not easy to get NLP to scale to the level of IR (as demonstrated by Google). IR is based on extensive *precomputation*—the creation of various kinds of indexes—that are then looked up. In principle, one could preprocess text, introduce markup (this is typically done using XML), and then index the markup. For some reason, the indexing problem has been underemphasized in academic research—though some commercial outfits have addressed the problem in specific contexts.

  The practical issue with LP-based indexing is that NLP goes much further than IR in trying to extract "meaning" from text. Psychology researchers studying both humans and animals have long known that meaning is very selective to the problem that you are currently addressing.

  - The brain has the capability of ignoring inputs that it does not care about at the moment—if you are hunting for a friend in a crowd, for example, you focus on faces, ignoring colors and the scenery.
  - Similarly, when you are skimming a research article or book, you are interested in where (and if) it addresses the problem on which you currently want more information. You will slow down when you encounter what you want and skim rapidly past material you consider irrelevant.

- Also, you may refer to a book more than once, looking for different information each time. If we did not follow such strategies, we would be drowned in data.

NLP must work similarly. Trying to annotate every word in a document is a fool's errand. For example, parts of speech are utilized in parsing a sentence, but indexing them as such is not very useful: the typical user could care less about identifying instances of verbs used as nouns. In general, information-extraction methods zero in on specific domain-relevant categories, such as mention of diseases, medications, therapies, and procedures during NLP of medical text. Obviously, identification of such terms is thesaurus-driven. Indexing of recognized concepts is obviously useful, but this task that has more or less been subsumed by IR technology. It is also useful to index negation or "hedging" (eg, "cannot be ruled out," "not unlikely," "possible") and tag the concept being hedged or negated. One might want to index change and its direction, flagging terms like "increase," "worsened," and "reduced" again tagging the concept referred to.

Beyond this, however, the nature of the annotation (and its indexing) would have to be highly subproblem-specific. For example, determining the temporal sequence of events is an interesting exercise. Equally interesting is *anaphora resolution*—determining which person or thing is referred to by a "he," "she," or "it," or recognizing part–whole relationships, such as a city within a state. But would you want to index this information? The whole objective of indexing is to improve performance when the same kind of query is asked repeatedly, and I can't think of many situations, at least in the biomedical domain, where users might demand this capability. (Things might be different in other areas, such as the legal domain.)

In any case, IR software not only shows you the hits, it also performs keyword-in-context (KWIC) display, where the matching phrase is displayed along with the text surrounding it (eg, the sentence before, including, and after). One hardly ever uses a world like "it" to refer to something in the previous paragraph; the subject is reintroduced by name. In other words, the referred entity is close by. So anaphora resolution may be less important in practice, because you allow humans to do the resolution mentally.

- The one NLP subproblem that has been engineered to scale is, of course, speech-to-text conversion, which employs techniques like HMMs. Question-answering systems or voice commands (like Apple's Siri, which utilizes Nuance Corporation's technology) work in relatively limited circumstances, as most iPhone users have discovered. The voice commands are very basic (eg, "Call 9-1-1"): the question-answering systems themselves use a Google-type information retrieval algorithm—sensing keywords in the query—to search large volumes of data and return the most "relevant" answer.

The IBM Watson hardware–software system did this, winning a Jeopardy contest in 2011. However, there are concerns that Watson was basically highly tailored to a single problem—answering "Jeopardy"-type questions—and very basic ones at that, where the question was framed in a particular way, and the answer could have been retrieved by a human being using Google. For a discussion of this issue, see the article by Ken Jennings, one of the contestants [13], and the last section of Nadkarni et al. [14].

- The tools that enable nontechnically adept individuals to employ NLP for their own needs are, as of today, very few (and certainly none available as freeware or open source). The one commercial system that focuses on ease of use is Linguamatics' I2E software [15], which focuses on the biomedical area. This software leverages the extensive work done in the area of biomedical controlled vocabularies, and adds value in various ways, including incorporating IR technologies.

Setting up NLP in a domain like clinical medicine, where controlled vocabularies have long existed, requires considerable curation to create subsets that are actually useful for text indexing. There are concepts with names a dozen words long, which would almost never be encountered in text. Also, spurious synonyms (terms wrongly flagged as synonyms through indifferent curation) require weeding out: Demner-Fushman et al. describe work in this area [16]. In biomedicine, however, you at least have a starting point. For most other domains, the vocabularies and thesauri don't even exist, so you have to start from scratch.

4.10 FURTHER READING

Machine learning is very closely tied to data mining and big data—most books addressing the latter will reiterate data mining concepts. I'll therefore refer you to the last section of that chapter for details. For now (unless you want to turn to that chapter right away), just note that the data-mining texts make machine-related topics much more accessible to nontechnical readers. This is understandable because, to use an automotive analogy, these books teach you how to drive a car, while the machine-learning books teach you how to build a new gearbox or internal combustion engine. I'm guessing that most readers of this book also want to learn only the bare minimum of theory necessary to use the machine-learning toolset appropriately, and the best way to learn is by doing. The data mining chapter refers you to freeware tools that you can experiment with.

BIBLIOGRAPHY

[1] G. James, D. Witten, T. Hastie, R. Tibshirani, An Introduction to Statistical learning with Applications in R (Springer Series in Statistics). Springer, New York, NY, (2013).

[2] F. Provost, T. Fawcett, Data Science For Business: What You Need to Know About Data Mining and Data-Analytic Thinking, O'Reilly Media, Sebastopol, CA, (2013).

[3] N. Silver, The Signal and the Noise: Why so Many Predictions Fail—But Some Don't, Penguin Press, New York, NY, (2012).

[4] P.V. Joshi, Perpetual enigma. In: Perennial Fascination With All Things Tech. San Jose, CA; 2015.

[5] J.L. Bentley, Programming Pearls, 2nd ed., Addison-Wesley, Reading, MA, (1999).

[6] A.C. Bulhak, Postmodernism generator. In: Communications from Elsewhere; 1996.

[7] J. Stribling, D. Aguayo, M. Krohn SCIgen—an automatic CS paper generator. Available from: https://pdos.csail.mit.edu/archive/scigen/; 2008.

[8] S. Luntz, Journal accepts paper reading "get me off your f---ing mailing list". IFLScience 11/23/2014. Available from: http://www.iflscience.com/technology/journal-accepts-paper-reading-get-me-your-fucking-mailing-list; 2014.

[9] A. Ng, M.I. Jordan, On discriminative vs. generative classifiers: a comparison of logistic regression and naive Bayes, in: T.G. Dietterich, S. Becker, Z. Ghahramani (Eds.), Neural Information Processing Systems, Neural Information Processing Systems (NIPS), Vancouver, BC, Canada, 2001.

[10] G. Forden, False alarms in the nuclear age. NOVA: Military and Espionage. Available from: http://www.pbs.org/wgbh/nova/military/nuclear-false-alarms.html; 2001.

[11] Ö. Uzuner, B. South, S. Shen, S. DuVall, 2010 i2b2/VA challenge on concepts, assertions, and relations in clinical text, J. Am. Med. Inform. Assoc. 18 (5) (2011) 552–556.

[12] C. Manning, H. Schütze, Foundations of Statistical Natural Language Processing, MIT Press, Cambridge, MA, (1999).

[13] K. Jennings, Ken Jennings Op-Ed: 'Jeopardy!' champ says computer nemesis Watson had unfair advantages. Daily News February 17, 2011. Available from: http://www.nydailynews.com/opinion/ken-jennings-op-ed-jeopardy-champ-computer-nemesis-watson-unfair-advantages-article-1.139563; 2011.

[14] P.M. Nadkarni, L. Ohno-Machado, W.W. Chapman, Natural language processing: an introduction, J. Am. Med. Inform. Assoc. 18 (5) (2011) 544–551.

[15] Linguamatics Corporation. About I2E. Available from: http://www.linguamatics.com/products-services/about-i2e; 2015.

[16] D. Demner-Fushman, J.G. Mork, S.E. Shooshan, A.R. Aronson, UMLS content views appropriate for NLP processing of the biomedical literature vs. clinical text, J. Biomed. Inform. 43 (4) (2010) 587–594.

CHAPTER 5

Software for Patient Care Versus Software for Clinical Research Support: Similarities and Differences

5.1 INTRODUCTION

The electronic health record (EHR) is a complex and sophisticated software package. Software that captures clinical parameters during research studies—a clinical study data management system (CSDMS)—can be equally sophisticated: this is especially so of systems designed to capture data on an arbitrarily large number of studies, and which can be operated without requiring a large team of software developers. Superficially, it might seem that an EHR could be readily retrofitted to perform clinical data capture, but this is true only in a very limited set of circumstances.

> **At one point in the past, a well-known EHR vendor suggested in their documentation that their system could perform most of a CSDMS's functions. This was roughly equivalent to suggesting that a family sedan would be competitive in a Formula 1 racing event. This vendor recently got into the business of providing most the basic support for research studies—designating studies, assigning patients to studies, and allowing the design of study-specific alerts, and so forth. Once they realized how much CSDMSs actually did, this claim was retracted. The vendor now acknowledges that their product's functionality is ancillary to that provided by a full-fledged CSDMS.**

This doesn't mean that EHRs play no role in clinical research at all: their effective interoperation with CSDMSs can be vital to the smooth conduct of a long-term study. Knowing how to achieve interoperation separates the good clinical-research support teams from the merely average ones. The succeeding section is partly abstracted from Chapter 14 from my previous book on *Metadata-Driven Software Systems in Biomedicine* [1]. More than 4 years have passed since that book was written, however, and the landscape for both EHRs and CSDMSs has changed considerably: the following text reflects those changes, even though certain fundamentals remain.

At the outset, I should mention that the needs of clinical research are sufficiently diverse that there is no single clinical-research software package that meets all needs completely. As I'll discuss in the next chapter, CSDMSs are only one of several types of systems that are employed. The others include grant management systems and biospecimen inventory systems. Even among systems that store clinical parameters, some systems

Clinical Research Computing. http://dx.doi.org/10.1016/B978-0-12-803130-8.00005-1

specialize in data related to genetics, and others have specialized functionality related to doing phone interviews.

5.2 SIMILARITIES BETWEEN EHRS AND CSDMSs

5.2.1 Data model

Both EHRs and CSDMSs are designed to store data on any number of individuals. Both allow the user to record data on an arbitrary set of clinical parameters, including those that the vendor never heard of (eg, parameters discovered through leading-edge research). They do this using the Entity–Attribute–Value model described in chapter: Core Informatics Technologies: Data Storage.

Large-scale CSDMSs tend to employ an EAV model for clinical parameters almost exclusively: EHRs employ EAV much less than CSDMSs do. This is because certain parameters, such as Vital Signs, are recorded for all patients (ie, the data is *not* sparse), and so you have explicit columns for Height, Weight and Temperature. Where data is sparse (eg, lab-test results, medications administered or ordered, or diagnoses), EHRs employ special purpose tables, where *many* different values hang off a given entity-attribute pair. Thus, for a lab-test result, values include the date/time the lab test was performed, the ID of the technician performing the test, and whether the result is normal, low, or high. For a medication, one might record preparation strength, total quantity dispensed, dose schedule, and administration instructions.

EHRs, however, allow forms to be built to capture user-defined parameters. Epic calls this feature "flowsheets": these are often employed by inpatient nursing staff. You define custom parameters in a data dictionary, and then combine these parameters into a form.

5.2.2 Separation of transactional and query/analytic functions

Both EHRs and CSDMSs often need to support transactional operations, where data are added, deleted, or changed, for a large number of concurrent users. Running complex queries that could return unpredictably large volumes of data can slow down transaction performance unacceptably. Therefore, at regular intervals (eg, nightly), they typically replicate the data on to a read-only database (henceforth termed the *analytical database*) that is dedicated to running queries, which resides on separate hardware. While the queried data is slightly out of date, such a compromise is usually acceptable.

Depending on the technology employed, the copying may be *incremental* (ie, only changed data is transferred), or *full* (if the source system does not readily support the identification of changed data). The query-oriented database may use different technology compared to the transactional system: thus, the Epic EHR uses a nonrelational database engine (InterSystems Caché) for transactions, and a relational database (either Oracle or Microsoft SQL Server, depending on the customer's preferences) for query.

Caché's design was inspired by MUMPS, a language developed at Mass General in the 1960s—and originally employed by Epic, as well as by the Veterans

Administration for its own EHR. MUMPS itself is considered archaic technology today: that it has formed the basis for some pretty robust systems demonstrates that good design and responsiveness to the user's needs are usually more important than the technology employed *per se*. Even state-of-the-art technology can be abused to create an unusable system when employed by incompetents or vendors who refuse to listen to their customers.

The database technology supporting EHRs and CSDMSs has relatively rudimentary analytic capabilities—though, with offerings like Microsoft SQL Server Analytical Services, this is changing. You therefore usually need to get data out from these systems into your favorite statistical packages or data-mining software to do serious analyses.

We now consider the far more numerous differences.

5.3 EHRs ARE SPECIALIZED FOR CLINICAL CARE AND WORKUP

While CSDMSs and EHRs both deal with clinical data capture, their underlying database designs are significantly different from each other. These differences reflect their respective intended purposes.

- In clinical research, patients are *preselected* based on their previously determined diagnosis and associated traits such as comorbidities. By contrast, it takes time to diagnose a new patient, or new problems in an existing patient: EHRs facilitate diagnostic workflows.
- The visits between the patient and the healthcare provider are mostly inherently unpredictable and driven by the presence of ailments that need treatment. In a clinical study, the schedule of visits is mostly determined when the study is being designed.

5.4 CSDMSs: STUDY PARTICIPANTS (SUBJECTS) ARE NOT NECESSARILY PATIENTS

CSDMSs, unlike EHRs, are intended to support workflows that are completely unrelated to patient care. Subjects, unlike patients, are not necessarily sick, and may not be followed longitudinally—perfectly healthy people may be surveyed for their opinions just once, for example.

Patients must always be identified because we do not want to perform an intervention on the wrong patient. However, research subjects may be anonymous, depending on the study design. IRBs do not like personal health information (PHI) to be captured if the study design does not mandate it, as for purely retrospective analyses of data. Here, if one is doing a chart review, one would obviously need access to PHI to identify the patients on whom data is being extracted. However, the extracted data need not record this PHI any longer than is necessary to be able to go back to the patient's record if necessary.

In genetic studies, long-dead ancestors, about whom we may not even know anything besides their names, may be included as subjects merely to connect families of individuals who are still alive and increase the statistical power of an analysis.

The Icelandic company Decode Genetics used Iceland's meticulously maintained genealogy records, coupled with the electronic medical records of Iceland's national healthcare system, to create enormous family trees that included most of the country's present population in some of their analyses, obtaining insights into various genetic conditions that were previously impossible with smaller groups of individuals. The approach to getting this data without consent of the individuals was eventually killed off by a decision by Iceland's Supreme Court. Studies of populations such as the Amish or certain Native American tribes have been facilitated by genealogy sources to which individual communities' governing councils have granted access.

5.5 STUDY PROTOCOL: OVERVIEW

The concept of a *study* is central to CSDMSs. The various aspects of a study that determine its conduct, such as the objectives, subject selection criteria, the parameters that will be recorded, and the times when they will be sampled, comprise the *study protocol*.

The word "protocol" has multiple meanings in biomedicine. It can also refer to a standardized means of working up a patient with a particular condition, including recording history and physical examination findings, and ordering specific investigations. In cancer therapy, it refers to the choice of specific interventions—chemotherapy, radiation, surgery, immune-stimulants, hormones, etc.—administered individually or in combination, and the doses and timings of each.

The various aspects of the protocol definition include the following areas. I discuss the nonobvious aspects in depth later.

- *Administrative details*: The name of the study, a brief description of the objectives, the sponsor, the IRB protocol number, sponsor's grant number, total budget, expected number of patients to be recruited, etc. Certain software, known as *grant management systems,* are specialized toward addressing administrative and budgeting needs.
- *Configuration information*: Based on the nature of individual studies and the current user who is accessing the data, certain aspects of the user interface may be turned on or off.
- *Security concern*: The *users* who are authorized to access the study data, their role/s within the study (eg, primary investigator, coinvestigator, research staff, statistician, IT support, pharmacist, research coordinator). One must also designate their *permissions* with respect to data access—read versus change, access to identifiable data, and access to blinded data. The permissions can get highly granular: in double-blind studies, only the pharmacists are allowed to see what medication a patient is actually receiving.
- *The sites that are conducting the study*: Unlike EHRs, which operate within the firewall of a single institution, the sites may be geographically scattered across the nation or even across different continents. With modern Internet speeds, transcontinental clinical studies, where the data is hosted at a single central location, are quite feasible. Individual users may be associated with one or (rarely) more sites.

Many EHRs provide web access only in the most limited fashion (notably for patient portals), and workstations must often be extensively configured. Since they are not intended to be used beyond a single institution, this arrangement is viable, though expensive and tedious. Since large-scale research studies are multicentric, however, CSDMSs, need to support broad access across firewalls. This virtually mandates web-based access. Also, CSDMS software must be relatively easy to learn and operate at the end-user level, given the often high turnover in research staff, as discussed in Section 1.4.4.

In most studies, each subject who participates in a study will be assigned a unique participant ID. The participant ID comprises, for example, a combination of a Study Prefix, a Site Prefix, and a sequential number within that site. This allows unique identification of a data item as originating from a subject while maintaining privacy (the statistician does not need to see personal health information).

- *Subject eligibility criteria*: The criteria used to include (and exclude) subjects from participation. Some of these items can be gleaned from the electronic medical record, while others are based on questions that are asked directly of the subject. EHR query is, understandably, the most efficient way of narrowing the pool of eligible subjects.
- *The study calendar*: This is the sequence of planned encounters (*events*) between a subject and research staff, in terms of the time intervals between encounters and the data items that must be gathered at each encounter.
- *The patient calendar*: This comprises the schedule of planned visits for individual subjects and the planned workup for each visit. The patient calendar schedule is based on the study calendar.
- *Data entry forms* (also called *case report forms*): At each encounter, data is collected based on interviews or test results: the results are either manually entered or electronically imported into forms. Permissions to a form may be quite granular—some users cannot see a form, others can only view its contents but not edit it, and certain items on the form (eg, those requiring expert judgment to fill in) are off limits to all but a few.
- *Study arms*: Sometimes, based on specific characteristics (eg, gender, severity of condition, or how they respond to therapy), the study calendar and the data entry forms may vary across the subjects in a study. The "Arm" encapsulates this unit of variation. In a double-blind study with independent groups, each Arm may correspond to a treatment group (standard vs. test).
- *Projects*: A project is a higher-level group of related studies with some common characteristic (eg, all focused on the same disease condition). Sometimes, the same patient may be in more than one study at the same time. Common data on these subjects, such as demographics or screening information, is shared across all studies in the same project.

Even where EHRs attempt to support a particular aspect, they may fall short in one or more ways, as now discussed.

5.6 CONFIGURATION INFORMATION

Various aspects of the user interface can be preconfigured in CSDMSs depending on a given study's needs. Modern software (and not just for CSDMSs) will read the configuration information and change its user interface accordingly.

- Depending on the number of subjects to be recruited, it may be either more ergonomic to present the current list of subjects in a list, or (if they run into thousands) to allow search based on demographic criteria.

- Study IDs may be autogenerated or supplied by the user (in the latter case, they must be checked for uniqueness).
- For a study in which subjects are anonymous, fields that present PHI should be hidden. Similar hiding should occur if the current user is not authorized to inspect PHI. For double-blinded studies, nobody besides the pharmacist should see what medication a given patient is actually receiving.
- Certain studies may allow direct data capture from patients who can be given temporary logins and will see only their own data—and only those forms that have been designated as patient-enterable. For "branding" reasons—especially in patient-data-entry situations, individual studies may have specific graphic logos that appear on splash screens or certain parts of every form. (More on this shortly.)
- Certain capabilities of the software, such as biospecimen tracking, phone interviews (which require maintenance of callback queues, call history, etc.), automated randomization, and pedigree management, can be turned off if the study does not call for them.

EHRs, even the ones that have taken baby steps in supporting clinical research, rarely allow this level of customization.

5.7 RECRUITMENT AND ELIGIBILITY

The number of potential participants that one must contact to persuade them to enroll in a study is considerably greater than those who will actually enroll. Recruitment percentages of 10% or more are considered respectable. However, the CSDMS must record contact information about potential subjects, and also keep a record of all encounters, because the recruiting staff, who often work part-time, must be compensated for their efforts.

It is best if most of the eligibility criteria for a study can be recorded in a structured, *computable* form (as opposed to narrative prose) because it allows the possibility of querying the EHR for identifying potential subjects. Various approaches have been used, which tie the individual data elements used in the criteria to elements in controlled medical vocabularies, to which EHR data elements are also associated. The actual mechanics of querying vary.

- For most situations, one can directly query the analytical database supplied by the EHR vendor (mentioned earlier). Alternatively, data from the analytical database may be further extracted and curated (eg, by mapping elements to controlled vocabularies) into a clinical data mart. Further queries are run against the data mart. These approaches work best when patients do not need to be identified in real time—a delay of several days (or longer) between the patient presenting with the eligibility criteria and the identification process is acceptable.
- For certain studies, such as those involving patients in critical care or emergency situations, patients must be identified within minutes or hours of meeting the eligibility criteria so that they may be recruited as soon as possible. EHRs have several mechanisms for real-time decision support. The most widely used mechanism employ a vendor-provided rule engine that works (albeit not very efficiently) on the production data and generates alerts. (In the Epic™ EHR, these alerts are called "Best Practice Advisories.") The alerts can be composed through a graphical user interface, though using this productively requires some knowledge of how the

EHR data is organized. The actions effected by the alerts include sending messages (including pager messages or e-mails), generating orders, and so forth.

Warning: **Such alerting mechanisms can slow down the performance of the production system if used indiscriminately. Sluggish performance is undesirable in patient-care situations where delays in accessing an individual patient's data are unacceptable. I know of one well-known academic center that contracted with the vendor to implement many hundreds of alerts, which made the system so sluggish that they again contracted with the vendor to remove the vast majority of alerts. The vendor, instead of warning the customers about the performance degradation, got them coming and going, billing both times for their labor.**

However, automated electronic screening has its limitations.
- Many selection criteria, especially those in psychiatry studies, are highly complex and may not correspond to controlled-vocabulary codes except at a very coarse degree of granularity [2].
- *Ineligibility* criteria are unknown at the time of the patient's last encounter in the EHR, which may have occurred sometime before the current date. Patients do not visit the hospital when they are well, unless for a scheduled follow-up. Thus, we may not know whether a female patient is currently pregnant (which would exclude participation in any studies involving administration of experimental drugs).
- Some criteria require additional tests to be performed through screening procedures. For example, for a particular investigational chemotherapeutic agent in specific cancers, the presence of brain metastasis may exclude them; therefore subjects with neurological symptoms must undergo a CT scan/MRI of the brain.

CSDMSs must, however, record the eligibility-screening data of all potential subjects, including those that failed screening, if only because the sponsor must reimburse screening costs, which must therefore be documented.

5.8 STUDY CALENDAR

Most longitudinal research studies are conducted in ambulatory (outpatient) settings, if only because continued hospitalization is prohibitively expensive and inappropriate to most conditions of research interest. Consequently, the vast majority of visits are *scheduled*. (Unscheduled visits may occur occasionally, eg, if the subject shows up with a sudden adverse event or illness, which may or may not be due to the experimental intervention.)

The planned timetable of scheduled visits, relative (typically) to the first visit, is called the *study calendar*. More accurately, *the study calendar is the set of time points or events that are critical to workflow in the study*. Not all of these time points may involve a physical visit or even involve a subject/patient. Also, several time points may correspond to a single physical visit.

The timetable is study-specific and can vary from a single "virtual" visit (eg, a one-time web-based survey) to multiple events minutes to hours apart corresponding to a single physical visit (eg, a pharmacokinetic study), to visits over a timespan of one or more decades (eg, the US National Children's Study). Events not related to patient visits may include mailing letters or work preparatory to a subject visit (eg, ordering of supplies, reservation of facilities, and personnel for imaging investigations).

5.8.1 Patient calendar

Understandably, all subjects are not recruited at the same time: they "trickle in": in long-term studies, some subjects have completed a year or two while others are just beginning. Even if all of them were immediately available, the limited number of research staff could not handle them all at once if they showed up on the same day for a specific follow-up visit. Therefore, visits need to be staggered so that the staff workload is evenly distributed.

A *subject calendar* refers to the study calendar *as it applies to an individual patient*. It can be planned (ie, computed) in advance and provided to the subject as a *schedule of appointments*, with some leeway negotiated between the patient and research staff based upon public holidays and upon the subject's convenience.

One of the events, either the date of enrollment or the baseline visit, is designated as a "time zero" for a given patient, and all proposed future events for that patient are computed is typically specified from the "time-zero" event.

The schedule that is agreed upon and handed to the patient, and backed up during the course of the study by notifications through e-mail, letter, or phone, will also contain information about what each visit will entail. It will also specify what is required of the subject for a particular visit (eg, an overnight fast). Each event has a researcher-designated slack (*window*) (eg, the "1-year follow-up" event may be allowed to occur at between 11 and 13 months. This way, if the patient misses an appointment or notifies the research staff later about unavailability for a particular visit, that visit can be rescheduled as long as it falls within the window.

Specific actions—interviews, administration of therapy, special investigations—are performed at specific events. Expensive or dangerous interventions are performed much less frequently than cheap or routine ones. The research protocol must define which actions apply to which events, and the software must *enforce* this by detecting the current subject's present event and preventing against inapplicable actions. Thus, if the study involves an MRI at baseline and at the 6-month time point, "MRI" must not be presented as an option in the user interface when a given patient is about to show up for the 3-month visit.

The CSDMS should also provide research staff with notifications about impending visits, with details of what is planned for each visit so that the appropriate workflow (eg, scheduling of use of a scarce resource) can be planned, as well as reminders to subjects about the same. Sometimes, despite all efforts, the subject may miss a visit or show up much later, beyond the window for that visit: missed and delayed visits should therefore be tracked. (Data for the delayed visit may not be usable if it falls outside that event's time window.)

An additional benefit of employing the study calendar is that the data of all subjects can be aligned based on the study events so that data for the same event can be aggregated across subjects even if the chronological dates vary—as long as they fall within the respective event windows. This dramatically simplifies data analysis, allowing, for example, measurement of the time-course of improvement after a therapeutic intervention. By contrast, analysis of EHR data, where patients come in when they fall sick and may have greatly varying lengths of hospital stay, is much more complicated.

5.8.2 Differences between EHR and CSDMS data capture

To standardize data capture and minimize variation in care between an experienced practitioner and a new intern, EHRs make heavy use of electronic disease-specific protocols for patient workup. Standard forms are employed that present the significant history and physical exam findings, along with predefined sets of orders. This way, an inexperienced practitioner does not have to think or try to recall what must be done: the desired plan of action is already laid out (though additional information can also be volunteered as narrative prose).

> **Systems such as Epic allow institutions to create their own templates (called "SmartForms") for individual disease conditions. While data entry relies heavily on choices based on Yes/No responses or pull-down lists, the output is stored as text. However, for a given template, the output is highly predictable in that natural-language-processing programs can "cheat" and achieve high accuracy by relying on knowledge about a given template's structure, with the caveat that a custom program has to be written for each template in use. The number of templates in use can run into the hundreds.**

5.8.2.1 Use of structured data versus narrative prose

However, there is a catch. It is often the case that until a patient's condition/s are diagnosed (this is often the case in primary care or emergency medicine), we may not know which protocol to employ. Since the patient's complaints, history, and physical exam findings must still be recorded until the diagnosis is made, there is a significant reliance on narrative text, which alone can provide the necessary flexibility. (While templates may exist, they are necessarily crude and may do little more than divide the record into sections.) Unfortunately, narrative text is significantly harder and more laborious to analyze because of all the varied ways in which the same concept can be expressed.

In a clinical-research study, fortunately, this situation does not arise because subjects are preselected for a disease condition, and the data elements to be captured must be specified even before the research protocol is approved by the IRB. Therefore one can capture structured data right from the outset, with occasional "comment" fields to record information volunteered by the subject that may not fit into the predefined fields. Screens that maximize the use of choices (through pull-down lists, check boxes, or radio-buttons), especially if they use sensible defaults, are much faster to fill than entering the same information textually.

5.8.2.2 Granularity of data capture

Research data is also far more finely granular than EHR data. Research into a cohort of subjects with a specific condition can go into details that would simply be too time consuming to elicit and record during routine clinical care. Many psychiatric questionnaires, for example, such as the 550+ item Minnesota Multiple Personality Inventory, have been designed specifically for research.

Because the amount of data entered may be voluminous, CSDMSs have defined the state of the art in interactive data validation, as well as other aspects of electronic form operation that would be overkill in clinical care.

For example, some research designs [3,4] require a subject (or tissue) to be evaluated by more than one research team member, to quantify interobserver variability or agreement. Issues of agreement are important for research, especially when a new form/questionnaire is being developed and evaluated, but one rarely has the luxury of being able to do this in pure clinical-care settings.

5.8.2.3 Form reuse issues

Given the tedium of developing detailed electronic forms, it is desirable to maximize reuse of a form for multiple studies. The following capabilities now become essential.

- Before creating a new form, one must be able to search the form repository to determine whether an existing form could be reused instead. Search capabilities must therefore be powerful and flexible, as well as easy to use. One must similarly be able to search for individual items that can be reused across multiple forms.
- Forms must allow *study-specific customization*. Multiple versions of the Hamilton Depression Rating Scale are employed in psychiatric research, for example. One would like to define a superset of questions from which a subset could be chosen for a specific purpose. The most efficacious way to do this is for the software to allow extraneous questions to be hidden. Dynamic hiding, with reformatting of the form, is straightforward with web technologies, and much more complicated without them.

Unfortunately, in most EHRs (as well as an unacceptable proportion of CSDMSs), both of the previously mentioned capabilities are rudimentary. (Epic's flowsheet-management facilities, notably, are currently abysmal.) Consequently, the same parameter tends to be defined dozens of times with different names. Institutions with 20,000+ flowsheet elements representing only a modest fraction of that in unique concepts are commonplace: query and cleansing of such data is a major challenge that keeps data-warehouse teams fully employed.

The situation is even worse with forms (which are aggregates of individual parameters). Because the nonweb technology employed for flowsheet display does not readily allow hiding of nonrelevant elements, numerous variants of the same form may be created with one or another parameter omitted. The content of such repositories can be a poster child for an "uncontrolled vocabulary."

> To be fair, even if the repository management and customization facilities were top notch, the end result may still be chaotic. The curator of the repository

needs to have a cataloger/librarian mentality that borders on obsessive-compulsive to prevent semantic duplicates from polluting the repository. Such personality profiles are uncommon. In many organizations, the folks managing the repository are often nurses or data-entry clerks who have been taught to use the software in a hurry, but who have not had the idea of reuse (which ultimately facilitates analyzability of the data) drilled into them.

To complicate matters, senior IT management often doesn't understand the issues themselves. Consequently, defining data elements is considered a menial task that is often delegated to people who are not considered good enough to do other jobs—such as patient care, in the case of nurses—but who are too much trouble to fire. I've seen at least a couple of people, in different organizations, who exemplify one of Homer Simpson's dictums: "If you don't like your job, you don't strike, you just go in every day and do it really half-assed. That's the American way."

5.8.2.4 Subject/patient-entered data

Many CSDMSs, notably REDCap [5] (created by Paul Harris's group at Vanderbilt), began as survey tools that allowed subjects to self-enter questionnaires, and had many workflow features that facilitated this task. EHRs initially lacked this feature: partly through federal government initiatives that require that patients have access to their own data, EHRs now support "patient portals," where patients can look at their test results, schedule appointments, contact their provider if necessary, renew their prescriptions, and so on.

This capability has been retrofitted to support patient surveys. Epic, for example, presently allows questionnaires to be filled in through their patient portal (MyChart™). The process is triggered by an explicit e-mail message from the healthcare team to the patient, who then logs into the portal, navigates to the Questionnaires section, sees what forms are available, and fills in the necessary questionnaires.

This workflow is adequate for the occasional survey, but requires the manual intervention of IT staff (or the implementation of a specific EHR alert) to "push" the questionnaire to specific patients. It is less suitable for capture of data on a regular schedule, such as patient diaries that may be updated at a very high frequency (eg, daily). Here, it is preferable for the patient to log in directly, the workflow being driven directly from the study protocol. Further, the EHR technology lags considerably behind that of CSDMSs: features such as complex interactive validation or dynamic hiding of fields that don't apply (based on responses to previous questions) are not currently implemented.

Many self-entered electronic research forms, as employed in the PROMIS consortium [6], also employ computerized adaptive testing (CAT) [7], where the order of questions may be changed randomly, or the questions themselves vary based on how the patient has answered previous questions: no EHR currently supports CAT capability, because it so rarely applies.

5.8.3 Quality-control considerations

Many clinical-research studies (eg, retrospective analyses) do not require real-time data capture. In the past because of lack of electronic access, data was often captured on paper and then transcribed later into an electronic system, using an approach such as double-data entry (DDE) to minimize the chance of transcription errors.

DDE may have made sense in the punched-card era, when a single wrong/missed keystroke required a card to be discarded, repunched, and inserted in the correct place in a card stack, and the job rerun. However, in clinical research, the major source of error is not transcription but error at the source. Recording source data on paper provides no consistency checks at all, and DDE transcription of potentially inconsistent source data is, effectively, putting lipstick on a pig. See the article by Day et al. [8].

Until a decade ago, DDE was still prevalent in research settings, and many CSDMSs felt obligated to support it. This reflected a delusion that research data was somehow more critical than regular clinical-care data. Even though a wrong lab-test value can result in a chain of erroneous decision making that might kill a patient, anyone making a suggestion that high-priced doctors or nurses double-enter data runs the risk of summary beheading. Data originating from laboratory instruments is typically electronically bulk imported into the EHR using laboratory information systems as an intermediary.

Today, with the wide availability and affordability of mobile technology, there is absolutely no reason to use DDE. The preferred approach is to maximize the extent of *interactive* validation by using ergonomic aids (eg, selection from lists, using software to determine a 9-digit zip code from a combination of street address, city, and state, etc.), followed by random audits of subsets of data to look for outliers. Also, the entire form need not be audited (eg, it does not matter if phrases in comments are misspelled). Attention should be focused on those elements that are critical for analysis.

In the (increasingly rare) situations where offline transcription is unavoidable, it should be as little delayed from the original encounter as possible (eg, not more than 4 days later [9]) so that they can be validated right away. Data errors can be corrected only by querying the source document's human originator—but the originator will recall the encounter only if it is very recent.

5.9 MULTIINSTITUTIONAL OR MULTINATIONAL RESEARCH SCENARIOS

5.9.1 Site restriction

When multisite studies are conducted, the CSDMS must often accommodate situations that would never be permitted to arise in clinical care. Thus, the research consortium may include investigators at different sites who are professional rivals, and who must not be allowed to look at each other's subjects' data. In other words, most users need to be *site restricted*.

Investigators within a consortium cannot always be trusted to put their rivalries aside for the greater good, especially if the Federal Program Officer responsible for the consortium is insufficiently assertive. Sometimes, a "Chinese Warlord" situation arises where such fights spill into the open. I've known one initiative funded by the National Cancer Institute that was single-handedly rendered ineffective by a researcher who kept bickering with the others.

5.9.2 Optimizing software design effort

For transinstitutional studies, web-based solutions are virtually mandated. Nonweb solutions that require setup of individual workstations, with license keys installed on each, are unworkable. (In contrast, many EHRs employ traditional "fat" client-to-database server access—where a significant amount of computation is delegated to the desktop machine—or remote terminal-based access.) A web-based solution has the advantage that software development for the consortium can be centralized.

If each site developed its own data-capture method, it would be a full-employment scheme for IT personnel, but apart from the greatly expanded informatics budget, duplicate development would predispose to divergence in the definitions of data elements, making eventual consolidation of the data prior to analysis needlessly challenging. Research sponsors understandably look askance at such work duplication.

5.9.3 Cultural/language issues: localization

A challenge with multinational sites is when individual sites use different languages. Sites like Amazon, which have a multinational presence, have solved this problem using *software localization*, where the content is tailored to a specific language and culture without having to completely rework the software code. (Look up www.amazon.es and www.amazon.cn to look at the Spanish and Chinese versions.). Localization issues also include how dates and numbers are presented and the direction in which text flows.

Cross-cultural pitfalls are well known in the business world. Kotler and Armstrong's popular text on marketing [10] has a hilarious section on misunderstandings that have arisen due to cultural issues, such as product names that turned out to be vulgar slang or derogatory terms in other languages. There was also the case of a detergent advertisement in three panels: the left showing dirty laundry, the middle panel showing the washing, and the right showing spotlessly clean fabrics. Unfortunately, this ad was aired without change in the Middle East—Arabic and Hebrew are written and read from right to left.

Localization methodology was originally pioneered by Apple for their Macintosh platform in 1984 and copied subsequently by other platforms. Web technologies such as HTML5 and Cascading Style Sheets support localization. The secret to localization methodology is to avoid hard-coding text and icons containing text (eg, captions, prompts, banners) in program code, and instead employing language-specific "resources" that are stored on disk and referred to in the code by symbolic names.

Amazon prefers to employ separate websites: this approach makes sense because the content that it presents (eg, lists of top-selling items) may be culture specific. Modern web servers can sense the default language of the connecting web browser and load the appropriate text resources accordingly on demand. This is called *dynamic (ie, on demand) localization*. For CSDMSs, the resources would include the captions of clinical parameters and within forms (including headings), online help and the text of items based on choices (eg, pull-downs or radio buttons). Resources must exist for each language that you intend to support. Certain development environments (Microsoft Visual Studio) and frameworks (eg, Microsoft ASP.NET) make localization significantly simpler, and Visual Studio can even be configured to red flag hard-coded text within program code, which would interfere with localization.

EHRs rarely allow dynamic localization. Within an organization (which is less geographically scattered than a multinational consortium), a "majority language" is agreed upon when the software is purchased. I can imagine, however, that in parts of the United States where Spanish is used extensively, such as Puerto Rico or southern Texas, or the French-speaking parts of Canada, a dual-language feature might be desirable. This is significantly easier if the software is web-based, which many EHRs, unfortunately, are not.

BIBLIOGRAPHY

[1] P.M. Nadkarni, Metadata-Driven Software Systems in Biomedicine: Designing Systems That Can Adapt to Changing Knowledge, Springer, (2011).
[2] R.L. Richesson, P. Nadkarni, Data standards for clinical research data collection forms: current status and challenges, J. Am. Med. Inform. Assoc. 18 (3) (2011) 341–346.
[3] S.S. Thwin, K.M. Clough-Gorr, M.C. McCarty, T.L. Lash, S.H. Alford, D.S. Buist, S.M. Enger, et al. Automated inter-rater reliability assessment and electronic data collection in a multi-center breast cancer study, BMC Med. Res. Methodol. 7 (2007) 23.
[4] J. Van den Broeck, M. Mackay, N. Mpontshane, L.A. Kany Kany, M. Chhagan, M. Bennish, Maintaining data integrity in a rural clinical trial, Clin. Trials. 4 (5) (2007) 572–582.
[5] P.A. Harris, R. Taylor, R. Thielke, J. Payne, N. Gonzalez, J.G. Conde, Research electronic data capture (REDCap)—a metadata-driven methodology and workflow process for providing translational research informatics support, J. Biomed. Inform. 42 (2) (2009) 377–381.
[6] D. Cella, W. Riley, A. Stone, N. Rothrock, B. Reeve, S. Yount, D. Amtmann, et al. The Patient-Reported Outcomes Measurement Information System (PROMIS) developed and tested its first wave of adult self-reported health outcome item banks: 2005–2008, J. Clin. Epidemiol. 63 (11) (2010) 1179–1194.
[7] Wikipedia. Computerized adaptive testing. Available from: en.wikipedia.org/wiki/ Computerized_adaptive_testing, 2010.
[8] S. Day, P. Fayers, D.R. Harvey, Double data entry: what value, what price?, Control Clin. Trials. 19 (1) (1998) 15–24.
[9] S.R. Wisniewski, A.C. Leon, M.W. Otto, M.H. Trivedi, Prevention of missing data in clinical research studies, Biol. Psychiatry. 59 (11) (2006) 997–1000.
[10] P. Kotler, G. Armstrong, Principles of Marketing, Prentice-Hall, New York, NY, (2015).

CHAPTER 6

Clinical Research Information Systems: Using Electronic Health Records for Research

The term "clinical research information systems" is extremely broad, and refers to a family of software packages that manage various aspects of data generated during clinical research, including reporting based on the data. There are several subcategories of such systems.

- Biospecimen management systems
- Systems specialized for managing genome-related data on related individuals (pedigrees) and unrelated individuals (populations)
- Research grant management systems
- Clinical research workflow management systems, including
- Clinical study data management systems (CSDMSs), by far the most extensively used

Details of each are discussed shortly. I'll reserve clinical data management systems, which will be covered most extensively, for the last. At the outset, I'll state that many of the commercial systems play more than one of these roles, but no package serves all functions equally well. In some cases, functionality was grafted on with such little forethought that using that particular capability of the software package in question would be a grievous mistake. In other words, if you want to employ best-of-breed solutions, you will most probably end up using multiple packages.

Thus, BioFortis's Labmatrix™ [1] began as a biospecimen management system, but added clinical data management later. In my opinion, the clinical data component is below mediocre: trying to employ it for a longitudinal study will get you into trouble. Similarly, Progeny™ [2] began as a specialized package to support recording of genetics data—it lets you define alleles and genotypes, generates pedigree diagrams, and can also export data to several genetics-analysis programs. Biospecimen management was added later—it does a very good job for the workflow related to DNA samples—and clinical data management added still later. (This last component is also very limited, serving to do little more than record essential phenotyping traits for specific individuals.)

Clinical Research Computing. http://dx.doi.org/10.1016/B978-0-12-803130-8.00006-3

6.1 BIOSPECIMEN MANAGEMENT SYSTEMS

These systems support inventory and workflow (eg, specimen tracking) tasks related to biospecimens. Such systems are general purpose in that the specimens that are tracked do not necessarily have to come from research subjects—they could have been collected during the course of patient care (eg, after biopsy or surgery). Further, the samples do not even have to be human. Examples of such systems are Biofortis' Labmatrix, FreezerPro™ [3], and Freezerworks™ [4], all commercial, with the last also provided as a free, limited-capability version to academic institutions. CaTissue [5] is an open-source offering originally developed with support from the National Cancer Institute.

> **CaTissue had the reputation of being somewhat clumsily designed and less than intuitive to use. After the plug was pulled on NCI's CaBIG (Cancer Biomedical Informatics Grid) initiative [6], adoption seems to have declined greatly. Though it is now being supported by a commercial software development firm that has enhanced the package, support is not free anymore.**

6.1.1 Functions provided by biospecimen management systems

All biospecimens are not alike. Certain biospecimens, notably DNA and cancer cell cultures, are potentially renewable resources: DNA can be cloned, and cancer cells are potentially immortal, multiplying indefinitely under the right conditions.

You must track several details about a biospecimen:

- *Basic aspects* about a biospecimen include origin, specimen type, current quantity/volume, and "transactions" where quantities of the specimen are consumed. Liquid samples may be sent for analysis, cell cultures and tissue specimens may be sent to collaborators, or even sold commercially to requesters.
- *The "daughter" biospecimens that have been derived from it*, either by simple dilution in laboratory reagents (aliquots), or in the case of living cell lines, through subculture. Many cells cannot be subcultured/replicated indefinitely, even if the cells continue to live, because DNA (and cells containing DNA) can mutate. With a large enough number of mutations, the current generation of a cell line may differ quite significantly from the initial sample.

> **Certain preparations, such as yeast artificial chromosomes (human chromosomal segments ligated to yeast DNA and then grown inside yeast cells, where they multiplied so that they could be extracted later for genomic characterization) were prone to a rearrangement of the human DNA, called "chimerization." Here, noncontiguous segments of human chromosomes (sometimes, different chromosomes) shuffled around and combined. The order of genes was thus scrambled and different from their true order. Chimeras were so frequent that real progress in mapping the human genome was made only when YACs were abandoned in favor of bacterial artificial chromosomes (BACs), where the incidence of chimeras was negligible.**

- *Storage location*: We do not just track the present location, but keep a complete history of all previous locations very precisely, down to the rack/box level and location within a box.

Location has become critically important since the possibility of contamination from neighboring biospecimens was realized several decades ago. A particular cervical cancer cell line, called HeLa cells after the patient (Henrietta Lacks) from whom they were originally collected, proved to be ultra-invasive, the cell line equivalent of a weed. Even the few HeLa cells that were carried in an aerosol and happened to land in another tissue culture took it over, soon outcompeting the original cells in the culture. Scientific experiments that had been assumed to be performed on other cells turned out to be performed on HeLa cells, invalidating many decades' worth of scientific results.

Historically, HeLa cells were the first ever line of "immortal" cells, so-called because, unlike previously cultured human cells, they did not die after a few divisions, but kept multiplying. Among other things, these cells were used to grow polioviruses for the development of the polio vaccine. The full story is in Rebecca Skloot's book *The Immortal Life of Henrietta Lacks* [7].

6.2 GRANTS MANAGEMENT SYSTEMS

The grants administration staff in major academic institutions have to juggle a large number of research studies. Each study has a different "status." A study may transition from one status to another, as illustrated next.

- At the earliest possible stage, the investigator/s have conceived the study and applied for funding, but the intended financial sponsor (who may be a government agency, a charitable nonprofit, a commercial organization, or an internal institutional source) may not have approved the study.
- Next, the study may have been approved—for example, it may have received a very favorable priority score. However, funding may not have been received.
- Funding may have arrived—or a date for the funding to arrive has been promised by the sponsor. However, one or more regulatory hurdles have to be cleared. For example, for an investigation of a new drug in humans, an application must be made to the Food and Drugs Administration (in the United States) or equivalent regulatory body elsewhere.
- For a large study, one may also have to commence hiring individuals with specialized expertise if they are not already available.
  Approval must also be sought from the Institutional Review Board (sometimes called the Human Investigations Committee) within the organization before participants/subjects can be actively recruited.
- Once a research team has been assembled and all approvals obtained, subject recruitment can commence. If all goes well, the study continues until the target number of subjects has been enrolled and studied. As will be discussed shortly, patients typically "trickle in" to a study— that is, all the subjects who might be eligible to participate in the study are not immediately available.
- Sometimes, the study may be halted prematurely for one or more reasons, some good, others bad (from the perspective of the investigators). Many studies follow what is called a sequential design, an approach originally pioneered by Abraham Wald in industrial quality control [8] and by the British statistician Peter Armitage in clinical trials [9]. In a study comparing a conventional treatment or placebo with a novel treatment, one conceptually reanalyzes the data with each new subject's completion of the study. If at any point, the superiority of either

treatment is found to reach a statistical threshold of significance (the threshold is decided before the study begins), then one can stop the study without the significant expense of recruiting and studying all the subjects originally planned.

Grants management systems track study status and the associated financial aspects (expenditures), warn about impending deadlines, and so on.

6.3 CLINICAL RESEARCH WORKFLOW SUPPORT SYSTEMS

These systems serve to streamline individual functions related to research that are supported at an institutional level. A research center in an academic or commercial institution has multiple functional (and administrative) subunits, such as the Clinical Research Unit (where subjects may be housed for the duration of particular studies); analytical laboratories that perform specialized assays, both standard or custom (including genomic analyzes); the pharmacy that compounds and labels specific medications, including blinded medications for double-blind designs (see chapter: An Introduction to Clinical Research Concepts), nutrition services for patients on special diets, biostatistics and informatics support, training to junior investigators, patient outreach, and so on.

In academic centers, the "umbrella" group responsible for this support has typically received a Clinical and Translational Science Award (CTSA) from the National Institutes of Health. It makes sense to centralize such support functions. Some of these services are provided free—especially to support pilot work by junior investigators who lack their own funding—while others may be charged back to individual research grants (whose grant numbers must be tracked so as to ensure that any publications by the group acknowledge the CTSA group's contribution). No researcher is forced to use the CTSA group's services, but if the group develops a good reputation, researchers realize that it is cheaper to "outsource" internally to professionals who know what they are doing rather than try to do it themselves (or build their own teams).

Individual groups utilize software that is specific for their tasks. Thus, informatics support groups typically use issue/bug tracking software, source-code control systems, and development tools specific to particular platforms. Pharmacy Support and Clinical Research Units may use tools to help organize and track their various chores. In addition, there are certain tools, now discussed, used by all the support groups.

6.3.1 Service request management systems

When an investigator commences a study, the resources needed to carry out the study must be identified. Rather than forcing the researcher to contact individual support groups, it saves time if the researcher can access a central web-based resource—a service request management system (SRMS)—to list all the resources needed so that the requests for these resources can be routed into individual support groups.

The request may be preparatory to applying for funding, or after funding has been obtained. Ideally, the researcher should have worked closely with CTSA staff *before* funding was applied for, in order for the CTSA group to adequately budget for resources to carry out the requested work, in case it is extensive. In chapter: Supporting Clinical Research Computing: Technological and Nontechnological Considerations, I've provided tips to help you disincentivize researchers from contacting you after they receive funding.

Making a request is only the first step. A requester may not necessarily know *how much* of each resource is needed. It is the job of individual support groups to start separate conversations with the researcher, helping them clarify their request, so that the cost of each subtask can be estimated accurately: don't take the requester's estimate at face value. (As I've discussed in chapter: Supporting Clinical Research Computing: Technological and Nontechnological Considerations, Section 2.7, certain investigators may try to lowball their estimates: you don't want to be subsidizing them, putting in much more than your team is getting reimbursed for.)

Within the SRMS, each service is assigned a unit cost, with the unit and the cost depending on the nature of service. Some costs are variable, depending on the number of subjects to be enrolled in the study (eg, sample collections and analyzes) while other costs are fixed (eg, salaries of personnel). Biostatistics/informatics support is typically an hourly rate that kicks in after an initial free consult to determine the scope of the support.

The SRMS must log all requests—so that in case of future disputes, there is no ambiguity about what was requested and what was delivered. It may track their current status, if progress is logged by individual service providers, as well as the usage of various services, time to completion, whether the service was delivered satisfactorily, and so on. Usage tracking helps decision makers reallocate resources, depending on demand for individual services. Also, statistics regarding services provided are important in establishing the impact of the CTSA group on institutional research (and getting federal funding renewed).

While some commercial SRMS applications exist, they are not much better than the open-source offerings. A full-featured freeware offering that I recommend is SPARC (Services, Pricing, and Applications for Research Centers), developed by Andrew Cates and his colleagues at the Medical University of South Carolina [10]. It has a few user-interface idiosyncrasies (and someone on your team will have to know or learn Ruby, the language with which SPARC has been built), but the price can't be beaten, and the MUSC team continually improves the product.

One caveat: as with any open-source freeware package, you can't expect round-the-clock support; SPARC's developers have other responsibilities. You therefore have to take ownership of your installation and put in some effort into understanding its internals and customizing the system if necessary. SPARC provides some customization hooks, such as the ability to change the startup screen, but you will probably want to do much more.

6.4 CLINICAL STUDY DATA MANAGEMENT SYSTEMS

I've discussed the differences between CSDMSs and EHRs in the previous chapter. Open-source CSDMS offerings include REDCap (the most widely used) [11], Open-Clinica [12], and TrialDB [13]. Commercial CSDMSs such as Oracle's ClinTrial [14] and Oracle Clinical [15] can be priced in the millions of dollars. Even after the up-front investment, they need a team to maintain. Further, because they need to interface to other systems that exist in the institutions, custom development is still necessary after purchase.

In general, commercial CSDMSs are much less mature than EHRs. Incredible as it may seem, until a few years ago, Oracle Clinical's GUI interface was a direct port of a VT-100 terminal interface, with almost no consideration given to usability. Oracle now recommends ClinTrial (the result of an acquisition) to its new customers. Even for the usable packages, system-level documentation tends to lag: the customer's development team needs to spend significant effort figuring out the software's internals.

Almost all mainstream CSDMSs employ a single database schema to store an arbitrary number of studies: individual users access only the studies to which they have permission. One particular vendor employs a design that generates individual schemas for each study (and sets up/maintains each study for the customer), but I'd recommend this system only for small groups who don't have informatics/programming staff on site, and figure that paying the vendor continually for support is cheaper than hiring staff. (If the number of studies gets large enough, hiring staff, and employing a different system, is preferable by far.)

I've touched upon various aspects of CSDMSs in the previous chapter. Chapter 14 of my previous book on metadata-driven software systems [16] covers the process of setting up a study in detail, so I'll avoid repeating myself.

6.4.1 Study setup: the need for maximal structure

When setting up a study, you should maximize the proportion captured in structured form: while you may need to leave a "comments" field on individual forms to allow staff or patients to volunteer information that your original protocol doesn't capture, the research staff for the study must periodically monitor comment content, looking for phrases that arise repeatedly. Perhaps a certain side effect shows up very often that was not anticipated: you should promote this to a specific discrete question.

The importance of structure may seem obvious, but I've encountered naïve researchers who assume that you can treat a CSDMS like an EHR, entering arbitrary-sized progress notes and, at the end of the study, magically transform that data into something analyzable. Your job is to educate such folks and point out that a CSDMS is *not* an EHR. This approach may provide full employment to NLP programmers, but they will have to write a custom program tailored to the particular study (and this code will have to be discarded when a new study comes along). Structuring up front will spare you from future migraines.

6.4.2 Implementing REDCap

REDCap is currently the de facto standard for implementing investigator-initiated studies. As I mentioned in an earlier chapter, it is not intended for FDA-regulated studies. This is not because it wouldn't work for the same, but because it hasn't officially been "validated" by a third party: validation can cost hundreds of thousands of dollars, an investment that is hard to justify for a package that is given away freely. REDCap supports both studies where patients are tracked using personal health information (name, medical record number, etc.) and can also be employed as a data-capture tool for retrospective analyzes, or anonymous web-based surveys where PHI need not be recorded.

REDCap was originally developed to support point-in-time web-based surveys. Earlier versions of REDCap were limited in their ability to manage longitudinal study designs: the output it generated was not in first-normal form, so you had to spend some effort massaging it into a format that was analyzable. However, its latest version has remedied this defect, and if your group has used it for a long time, you are strongly advised to learn how to use the new features: your users will thank you for it.

One caveat about implementing REDCap: many CTSA support groups conduct a basic training session and then leave researchers to set up their own studies. This may work for simple, nonlongitudinal study designs. However, for longitudinal studies, the latest version requires you to think in terms of a normalized relational design. Few end-users have this kind of training, and your support staff should set up the study for these researchers, rather than let them dig themselves into a hole. The consequence of benign neglect is that *when* (not *if*) these users get stuck trying to pull data out and analyze it usefully, they will naturally contact you for help, and your support requirements (which could have been minimized through the appropriate stitch-in-time before the study began) will balloon. In any case, you need to consult with them before they commence the study, if only to ensure that the study will not have special support requirements that REDCap is not intended to handle.

> In one case I'm aware of, which occurred before REDCap was enhanced to support longitudinal study design, a research group assumed that REDCap would work and did not budget for informatics support. However, their study called for daily tracking of newborns admitted to the neonatal ICU with neonatal thrombocytopenia (where the blood platelets fall to dangerously low levels, with a risk of death or permanent disability due to internal bleeding). Some severely ill babies could potentially be in the NICU for 150+ days.
>
> This called for a longitudinal study design with a relational structure—at the time, REDCap forced you to designate columns like PlateletCount1, … PlateletCount150. The informatics group ended up designing a custom microcomputer database at the very last moment and had to eat the cost of design and implementation, since the person in charge of REDCap support had previously told the group, *without* looking at their study protocol, that REDCap would meet their needs.

REDCap's real-time reporting capabilities have been deliberately limited: you will typically export your data to an analytical package in order to do such reporting. Its data-export facilities are reasonably intuitive to work with. One problem with REDCap, reflecting its origins for survey data, is that, unlike some other CSDMSs (including the high-end commercial ones), it currently lacks a global repository of data element definitions that can be reused across multiple studies. A data element is *reusable* when its data type, name, description, maximum and minimum values or permissible set of values, upper and lower limit of normal, etc., are standardized, so that once defined, it can be used repeatedly across multiple studies.

Once you employ a global repository, it becomes a local vocabulary whose quality you want to maintain. (I discuss this issue later in chapter: Clinical Data Repositories: Warehouses, Registries, and the Use of Standards.) You want to control its contents and quality by using your own support staff—who are answerable to you—to define data elements, rather than let individual end-users (who are not) do so. If you let end-users perform data element definition, you will find the same data elements (eg, vitals, common laboratory tests, or history findings) defined over and over again. This is understandable, because with some systems, an individual user has no way to look at other people's study designs. Even if such a capability exists, many end-users may be too hurried to do so. While an individual user's needs may be met, study setup takes longer because each user has to define elements from scratch rather than pick rigorously defined items from a repository/library. Further, you can kiss all hope of doing integrated analyzes across multiple studies (which may be important for certain domains such as cancer) goodbye.

6.5 USING EHRs FOR RESEARCH

As I've discussed in the previous chapter, EHRs are primarily intended for clinical care, and their research capability was grafted on as an afterthought. However, they serve several purposes in research.

1. Identification of patients who are candidates for specific clinical studies. This has already been discussed in the previous chapter, Section 5.7.
2. They can be used to support research workflow for critically ill patients. Most important among these are alerting functions, which need to be used judiciously rather than indiscriminately, for reasons discussed in chapter: Software for Patient Care Versus Software for Clinical Research Support: Similarities and Differences, Section 5.7.
3. Appointment scheduling for research subjects is often most conveniently implemented using the EHR's capability, if it exists, or the hospital's patient-scheduling system, if it does not.
4. They may be employed instead of CSDMSs for particular kinds of clinical studies.
5. The relational-data-store part of an EHR is the basis for data warehouses or data marts, as well as the source for extraction of data related to specific patient groups.
   The last two functions will now be discussed in depth.

6.5.1 Using EHRs instead of CSDMSs for primary clinical-study data capture

The circumstances where EHRs are used instead of CSDMSs for primary capture of research data are relatively few.

- *Pragmatic clinical trials*, where studies combine clinical care with a research component.
- *Quality improvement initiatives* related to the employment of disease-specific data-capture protocols, through the use of standard templates (including standard order sets) to reduce variance among providers.
- *The personal health record (PHR) component* of the EHR can be deployed to capture patient-entered data (eg, patient-reported outcomes) through electronic forms that are sent to patients. Depending on the EHR, the PHR software may work well for this purpose, or it may not, and a lot depends on whether the individual patient has established a rapport with her/his healthcare provider, as well as a patient's capacity and interest in using electronic forms to communicate and participate in research, so that a message from the EHR is not treated as spam and ignored.

Being able to implement all of the aforementioned effectively depends critically on the level of collaboration between the hospital IT staff (who babysit the EHR) and research personnel (who generally report to a different authority). This theme is discussed in greater depth in chapter: Clinical Data Repositories: Warehouses, Registries, and the Use of Standards, Section 9.6.

6.5.1.1 Pragmatic clinical trials

Pragmatic trials, as stated in chapter: An Introduction to Clinical Research Concepts, are intended to evaluate therapeutic measures as they would be employed in actual practice rather than the rigidly controlled conditions of a traditional clinical trial. Therefore they must be done "on the cheap," with data created and captured, for the most part, by ordinary caregivers who are not specially trained or qualified to do research—and most of the time, are minimally (or not at all) reimbursed for their participation in the study.

Herein lays a quandary. If, for example, clinicians have been using narrative text to capture the patient's progress rather than recording parameters in structured form, the resulting data will be difficult or impossible to analyze without extensive natural language processing (NLP), which is not 100% accurate or manual abstraction (ditto). Therefore, the ability of the EHR to allow the design of custom interfaces, based on templates, should be utilized.

However, demanding *extensive* capture of structured data from busy, uncompensated providers may be met with open revolt, and so the custom data capture must be kept as brief as possible. Califf [17] cites a study where the custom data entry was limited to a single, faxed page (presumably, the providers were not using a centrally accessible EHR). If the study design unavoidably requires extensive custom data capture, the research proposal should budget for scribes to reduce the clinician's data-entry burden.

As I've stated in the previous chapter, EHR interfaces are generally less capable of custom data capture than those of the best CSDMSs, and they have the additional limitation of (typically) not being web-accessible across institutional boundaries. The result is that in a multicentric study, there is considerable redundancy of effort, with informatics personnel at each site having to set up custom data capture: this effort also requires coordination to ensure that all sites are capturing the necessary parameters in an identical fashion.

In pragmatic trials, however, EHRs have the advantage that every workstation in the areas where clinical care is being delivered will have access to the EHR so that providers are not forced to log on to a separate system just for particular patients who are part of a study. In multicentric pragmatic clinical trials, clinician/nurse convenience trumps the downside of redundant informatics effort, provided that the custom data-capture requirements are kept to a minimum. If the study is conducted entirely within a single institution, of course, the downside is even less.

6.5.1.2 Quality-improvement initiatives

The idea behind *structured data capture* for common ailments is that a new resident who is just fresh out of medical school should be able to capture the same data elements, and to the same level of detail, that an experienced and knowledgeable specialist in the disease would. This is possible by the EHR prompting for those elements in the user interface and presenting choices for each element where appropriate. If well-implemented, disease-specific templates reduce the burden on the clinician's memory and act exactly like checklists so that medical errors of omission become less likely, and variation in quality of care due to variance between healthcare providers' expertise is minimized.

The templates should capture the relevant findings for the disease (so that significant negative findings are also recorded explicitly). The clinician should be prompted to order standard tests as part of the workup, and even the medication orders standardized as far as possible, with more than one choice offered in case the first line of treatment is contraindicated for some reason in a particular patient.

Quality improvement may not seem to be research *prima facie*, but every such initiative has a research and evaluation component. Even if you set out with the best intentions, you can't always assume that template implementations, by the mere fact of their existence, will always make things better. If designed without careful thought, the consequences may be quite different from those intended.

One warning here: vendors very often provide disease-specific templates, and the temptation exists to reduce one's own work by using something readymade. Don't fall into this trap: always verify the template content with respect to current guidelines. Vendor templates may not be religiously updated when guidelines change. One well-known EHR's pediatric templates continued to recommend loperamide for pediatric diarrheas long after it was known that loperamide was contraindicated because of the risk of toxic megacolon, a potentially fatal condition.

Another use of EHRs for quality improvement is to perform simple alerts for health maintenance. An important consideration here is that if a system *knows* that some action is necessary, then the system should also *do* something about it if possible. For example, rather than simply nagging the physician that a patient's vaccinations are due, it should volunteer to put that patient into the appointment scheduling pipeline (or add the vaccination task to the next scheduled appointment), and only ask the physician to confirm. Such capability requires sophisticated integration if the EHR vendor and scheduling-software vendor are not the same.

6.5.2 Utilizing the EHR's relational data store for data extraction: some tips

As I've mentioned in the previous chapter, EHRs typically include a relational data store—a database whose contents are a functional clone of the transactional production system, which is typically regenerated every night and used for reporting and querying in order not to impair performance of the transactional system. While you can run queries against this system directly, I would strongly recommend against doing so. Instead, you first copy selected portions of this store—specifically, selected fields from selected tables—in batch to a separate database, the operational data store (ODS), discussed in chapter: Clinical Data Repositories: Warehouses, Registries, and the Use of Standards, Section 9.2 and run queries against the ODS instead. (Obviously, you would set up a nightly batch-extraction process.) The reasons for doing this are listed as follows.

- Most EHRs include thousands of tables, the vast majority of them completely irrelevant to the needs of research. You may not really care about the quantity of individual medications or bandages stored in inventory, for example, even though this information is critical to those who need to reorder supplies.
- Even within a table, there are many columns that are used for internal housekeeping by the transactional system, about which researchers could care less. If you use a quasigraphical tool like SQL Server Management Studio to compose your queries, you will have to wade through an endless list of tables—and huge field lists for a table, most of which you don't care about.
- Also, many EHR tables have a large number of "legacy" fields that were used a long time ago but are now left completely empty, with other fields doing the work. (The reason for this is not obvious, but may be traced to the revolving-door nature of employment in certain EHR vendor shops. Certain tables were created for one purpose and get orphaned when the person who created leaves: someone else creates a table with an almost identical function. This obviously reflects a failure of project management, but then, nobody claims that EHRs represent the leading edge in software practices.)

Unfortunately, the legacy fields may still be left lying around as a trap for the new hires in your group who may fetch their contents not knowing that these fields are useless (and omitting to retrieve the similarly named fields where the needed data is currently stored). By creating a reduced set of the fields that you really need, the essential aspects of the system are more easily learned by your team and the chance of errors is reduced so that you serve your customers more efficiently.

The list of tables and fields that you pull is not static. As your knowledge of the EHR evolves—and believe me, no one tries to remember the full details of 12,000+ tables—you may find that you may need to pull additional fields, or additional tables, that you had ignored earlier, because some researcher comes to you with a special request and your current set of tables/fields lacks the necessary information. My own group had to revise our extract because of a study where a researcher wanted inpatient bed information—which we had not extracted earlier—in order to randomize patients by bed to a standard intervention versus a novel intervention. Another researcher needed the billed charge (in dollars) of individual procedures, which we had previously ignored when we were extracting billed-procedure data.

- You can create views that combine commonly used tables. You are not allowed to create views on the relational data store—if you did, they would get wiped out the next day when the store is regenerated. You can then create standard data extraction templates for various purposes (eg, extracting medication-related information for a cohort of patients, getting associated diagnoses, etc.) based on these views, which you can reuse repeatedly to streamline your data-extraction chores.
- You can add value by incorporating tables from other sources, such as medication hierarchies from drug knowledge bases (or even from subsets of the National Library of Medicine's Unified Medical Language System). Such hierarchies (other than the Dewey-decimal-type ICD-9 diagnosis codes) are rarely part of the EHR: they are useful for queries where a researcher would like to incorporate patient selection criteria based on these hierarchies, such as identifying all patients who are on any antiplatelet drug.

6.6 EFFECTIVE INTEROPERATION BETWEEN A CSDMS AND EHR-RELATED SOFTWARE

There are numerous instances in which data needs to be bulk-exported from an EHR into a CSDMS. For example, in a study on a novel therapeutic intervention, the CSDMS may be used for primary data capture to store a large number of custom parameters—study-specific questionnaires, novel laboratory parameters, and so forth. However, selected data that is already in the EHR on these patients—most commonly past diagnoses and selected standard laboratory tests—must be culled from the EHR. Most EHRs either incorporate a laboratory-test module or interact with the institution's laboratory-system software through an HL7 interface.

The reason for bulk importing is, obviously, to reduce tedium and errors on the part of the research staff caused by the need to manually transcribe information from one electronic system to another. Since there is rarely need for lab-test results to go into the CSDMS in real time, it is not too difficult to set up batch programs that access the EHR's laboratory-test-result tables for a specific set of patients and specific investigations. More important, since the EHR records date/times such as when the test was ordered, when the sample was collected, and when the assay was performed, you can take advantage of these timestamp fields to perform incremental extraction.

Depending on the CSDMS that you employ, the task of import may be easy or difficult. The open-source CSDMSs are, surprisingly, much easier to work with for automated data import chores than many commercial systems, whose internals are so poorly documented and opaque that the task of manual transcription, unpleasant as it is, may be preferable to the risk of corrupting data.

6.6.1 Transferring data from CSDMSs to EHRs

Moving data in the opposite direction—from a CSDMS to the production EHR—may be easy or very difficult. The latter circumstance applies when the application programming interface (API) of the EHR is extremely immature: this is unfortunately true of a disproportionately large proportion of EHRs, and the task of transfer has the same risks as skating on very thin ice.

This situation may improve. As part of Meaningful Use Phase 2, the Federal Government is requiring patients and their families be able to *download* their personal-health-record data electronically using an XML-based format called Blue Button (actually, Blue Button+). Blue Button format was originally devised through work at the Veterans Administration Medical Centers.

However, the federal website HealthIT.gov warns: "Please be aware, however, that some vendors and providers may be using the term 'Blue Button' consistent with the original, *narrower* technical set of requirements defined by the Veteran's Administration (VA), so use of the term does not automatically signify compliance with the full Meaningful Use Stage 2 requirements related to patient and family engagement." [18]. In other words, the functionality may be very limited: as a researcher, you have to experiment with your particular EHR to find out what it can or cannot do.

In principle, a vendor should also be able to *receive* a Blue Button message, parse it appropriately after authenticating its sender and contents, and move the data contained therein into the appropriate structured-data tables in the EHR (which are obviously vendor-specific). It is anybody's guess as to what extent this feature has been implemented with most EHRs: the Blue Button+ Implementation Guide was released only in 2013 [19].

Bulk import of structured data into EHRs is much trickier to implement than data export. To work reliably for lab tests, medications, and procedures recorded at a granular level, standard controlled-vocabulary codes must be used for each item, and an institution's EHR curators are responsible for mapping local vocabulary entries to these standard codes. As I discuss in chapter: Clinical Data Repositories: Warehouses, Registries, and the Use of Standards, Section 9.5, the state of most local EHR vocabularies tends to be abysmal, in large part because the people who maintain them don't understand the first thing about controlled vocabularies, and also because the EHR vendors don't provide tools to help assess, or maintain, vocabulary quality. In other words, at the moment, structured-data import is off the cards.

There is a proposal called "Green Button" [20], which is intended to let clinicians use *summarized* patient data (imported from external sources) for real-time medical decision-making. However, it is anybody's guess as to when and if such an idea will be implemented: usually, decision support requires highly granular, structured information rather than summarized information that might require custom NLP to process.

BIBLIOGRAPHY

[1] Biofortis Corporation. Labmatrix. Available from: www.biofortis.com/products/labmatrix/, 2014.

[2] Progeny Inc. Progeny genetic pedigree software. Available from: http://www.progenygenetics.com/, 2015.

[3] Ruro Inc. FreezerPro. Available from: www.ruro.com/freezerpro, 2011.

[4] Freezerworks.com. Freezerworks Unlimited. Available from: http://www.freezerworks.com/index.php/freezerworks-2015/, 2011.

[5] National Cancer Institute. CaTissue Suite. Available from: https://wiki.nci.nih.gov/display/caTissue/caTissue+Suite;jsessionid=DC1C41A7419089AE1F8E7D469816FB45, 2011.

[6] National Cancer Institute. Community Cancer Centers Program. Available from: http://ncccp.cancer.gov/Related/CABIG.htm, 2010.

[7] R. Skloot, The Immortal Life of Henrietta Lacks, Crown, New York, NY, (2010).

[8] A. Wald, Sequential tests of statistical hypotheses, Ann. Math. Stat. 16 (2) (1945) 117–186.

[9] P. Armitage, Sequential Medical Trials, 2nd ed., Blackwell Scientific Publications, Oxford, UK, (1975).

[10] Medical University of South Carolina. SPARC request: research made easy. Available from: https://sparc.musc.edu/, 2015.

[11] P.A. Harris, et al. Research electronic data capture (REDCap)—a metadata-driven methodology and workflow process for providing translational research informatics support, J. Biomed. Inform. 42 (2) (2009) 377–381.

[12] Akaza Research. Openclinica. Available from: https://community.openclinica.com/, 2011.

[13] P.M. Nadkarni, et al. Managing attribute-value clinical trials data using the ACT/DB client–server database system, J. Am. Med. Inform. Assoc. 5 (2) (1998) 139–151.

[14] Oracle Corporation. Phase Forward ClinTrial. Available from: http://www.phaseforward.com/, 2010.

[15] Oracle Corporation. Oracle Clinical. Redwood Shores, CA: Oracle Corporation, 1996.

[16] P.M. Nadkarni, Metadata-Driven Software Systems in Biomedicine, Springer, London, UK, (2011).

[17] R.M. Califf, Clinical trials, in: D. Robertson, G.H. Williams (Eds.), Clinical and Translational Science: Principles of Human Research, Academic Press, Los Angeles, CA, 2008.

[18] HealthIT.gov. Will using Blue Button satisfy the "patient and family" requirements of Meaningful use? Available from: https://www.healthit.gov/providers-professionals/faqs/will-using-blue-button-satisfy-patient-and-family-requirements-meaningf, 2013.

[19] BlueButtonPlus.org. Blue Button+ implementation guide. Available from: http://bluebuttonplus.org/, 2013.

[20] C.A. Longhurst, R.A. Harrington, N.H. Shah, A "green button" for using aggregate patient data at the point of care, Health Aff (Millwood) 33 (7) (2014) 1229–1235.

CHAPTER 7

Computer Security, Data Protection, and Privacy Issues

7.1 SECURITY BASICS

The necessity of securing communications existed long before the computer era. The most basic form of encryption, the Caesar cipher [1], where one letter of an alphabet, or a symbol, substitutes for another letter, existed since Roman times. Such *monoalphabetic* ciphers are fairly easy to break, relying on frequencies of letters, pairs of letters, and words in individual languages: the Sherlock Holmes short story, "The Adventure of the Dancing Men" [2] describes how Holmes cracks such a cipher.

More complex schemes involve the substitution rule changing with every letter (a *polyalphabetic* cipher). In the basic Vigenere polyalphabetic cipher, the alphabet cycles back to the original after every 26 letters (in English), thus making frequency analysis that much more difficult. *Transposition* ciphers change the order of letters and are typically combined with polyalphabetic substitution. *Block* ciphers, on which modern electronic methods of encryption are based, rely on substitution and transposition, but at the level of bits (1s and 0s) rather than "letters," and operating on blocks of bits at a time. In espionage, encryption is often accompanied by *steganography*, where the message is itself concealed—using invisible inks, microdots, or within JPEG images.

Concurrently, the science of *cryptanalysis*—the breaking of encryption—has also developed. Simon Singh's *The Code Book* [3] provides a fascinating overview of this field that is informative to both the layperson as well as the serious student. The only provably unbreakable code is the "one-time pad," a polyalphabetic cipher where the key, a copy of which is possessed by both the sender and the recipient, is randomly generated, and is as long as the message itself (and is never reused). This was actually employed by the KGB for manual encryption in the precomputer era, but is not feasible for use between two geographically separated individuals who interact for the first time unless the key is transmitted first over a "secure channel"—a way of transferring data that is resistant to both eavesdropping and tampering. Of course, no perfectly secure channels exist: the Internet, particularly, is not a secure channel.

In addition to being able to decrypt a message, an authorized recipient of the message must also be able to *authenticate* the message (ie, verify both that it came from the expected sender rather than an impersonator, and that the message itself was not altered).

Clinical Research Computing. http://dx.doi.org/10.1016/B978-0-12-803130-8.00007-5

Modern data-transmission techniques, which are performed by computer programs, deal with both issues where necessary.

7.1.1 Message/content integrity

Stored data is referred to "data at rest," while data being transmitted is "data in transit." In both circumstances, it is often important to ensure that it is not altered. This is accomplished by transforming the content (called a "message" in the transmission context) mathematically into a number called a *hash* (also called a *message digest* or *checksum*), which is much smaller than the message itself, using a mathematical transformation with two properties.

1. The chance of two different messages producing an identical checksum (a "collision") is vanishingly small.
2. The chance of reconstituting the original message from the hash is similarly miniscule. The only method of determining the original message would be "brute force"—generating all possible messages to see which produces the same result, a task that might take centuries of computer time.

The hashing technique currently recommended by NIST is called SHA-2 (SHA, Secure Hash Algorithm), the most secure variant of which generates a 512-bit hash, irrespective of the message's original length.

After a message is transmitted, the recipient computes the checksum and compares it with the transmitted checksum. If the two are identical, we have an assurance that its contents were not damaged or altered. In combination with public-key encryption, described shortly, the checksum is also used as the basis for digital signature and authentication (Section 7.1.3).

7.1.2 Encryption methods

Encryption methods similarly rely on mathematical principles where, without knowing the key (information that was applied to the message to encrypt it), decryption would take a prohibitive amount of computing resources. They fall into two categories: *symmetric* methods, where the same key is used for both encryption and decryption, and *asymmetric*, where different keys are used. The historical methods of encryption were all symmetric: the current NIST standard is Advanced Encryption Standard (AES) also called *Rijndael*, a play on the names of its Belgian inventors. (The "j" is pronounced as "y.")

Asymmetric encryption techniques are also called *public-key* techniques because one of the two keys used is *publicly* disclosed, while the other key is secret or *private*. They are much slower than symmetric techniques in terms of the amount of data encrypted or decrypted per unit time. Therefore they are used mainly for transmission of keys, which are then used for symmetric encryption over insecure channels, such as over the Internet. Such a mechanism is employed by Secure Sockets Layer (SSL), the protocol involved when you access a URL with the https://prefix, such as when transmitting credit card information to an e-commerce site.

The most well-known asymmetric encryption technique, RSA (named for its inventors, Ronald Rivest, Adi Shamir, and Leonard Adleman) is based on the fact if you take two large prime numbers (which are also kept private) and multiply them to yield a still larger number, factoring that large number into the original two primes is known to be a problem that can currently only be solved by brute force, and goes up exponentially with the size of the number. (Factoring a 232-digit number took the equivalent of 2000 continuous years of computation on a high-end workstation—it took 2 years on a cluster of thousands of machines.)

> **The prime factor problem is one of a family of mathematical problems called NP-hard problems (NP, nondeterministic polynomial). An example of an NP-hard problem is the famous "Traveling Salesman" problem—given a large number of cities and the distances between each, find the shortest route that will visit all cities, and visit each city only once. The time required to solve such problems goes up as the exponent or factorial of the problem size (eg, number of cities), and can rapidly become prohibitive computationally: the only way to tackle them is to be content with a less-than-perfect solution. In the case of the Traveling Salesman problem, there are relatively fast solutions that are guaranteed to, say, not be worse than 10% longer than the shortest route. For the prime-factor problem, a less than perfect solution is useless.**

> **In theory, if "quantum computers" ever get invented, they will solve NP-hard problems in polynomial time, which potentially works out to billions of times faster than existing approaches. However, the technical challenges to building a working quantum computer are formidable: existing proof-of-concept experiments have only solved "toy" problems.**

The Wikipedia article on public-key cryptography [4] provides an excellent explanation of how it can work robustly even in the presence of eavesdropping. In principle, you encrypt a message using the intended recipient's publicly available key: the message can only be decrypted by the recipient using the recipient's own private key. (This, of course, assumes that the recipient's private key has not been stolen in advance by the eavesdropper. This is why private keys are typically themselves secured by another layer of encryption. If an unencrypted key is discovered to be stolen, it can be *revoked*—the equivalent of closing a compromised bank account.)

7.1.3 Digital signatures

Signing a document digitally is employed by healthcare providers and researchers for patient data. Digital signatures have the following characteristics:
- *Integrity*: Subsequent alteration of a document's contents (eg, if a provider tries to cover up one's tracks after a patient experiences an adverse event) can be detected immediately.
- *Authentication*: We have a guarantee that it was signed by the person named on the document.
- *Nonrepudiation*: The signer cannot deny signing the document or claim that the signature was forged. This assumes, as stated earlier, that the signer's private key was not stolen by an intruder (eg, by employing malware that allows remote access and full control of the machine where the key resides).

The principle of creating a digital signature is to create a hash/message digest of the document (which ensures integrity) and then "sign" it by encrypting the hash with the signer's private key. The encrypted hash is appended to the document. The only way to determine the original hash is by decrypting the encrypted hash with the signer's public key (but not with anyone else's), which means that only the signer concerned could have signed it. If the document's contents were altered retrospectively, the altered document's computed hash (which can be readily determined by standard software) would be different from the recovered hash.

7.2 SPECIAL CONCERNS RELATED TO PERSONAL DATA

As opposed to content such as trade secrets, which must be protected in their entirety, data about people has the challenge that personal health information (PHI), such as the first and last names, date of birth, social security number, credit card information, and so on, require an extra layer of protection compared to other information on that individual—such as the items a customer ordered or the diseases that a patient suffers from.

Most electronic systems typically employ a surrogate ID to identify individuals—a "fake ID" that is typically a sequentially increasing number that is meaningless outside the database where it is employed. The PHI fields are localized to a single table in the database, which also employs the fake ID. If the PHI-containing fields are selectively encrypted, then even if an electronic intruder were able to download the database, she/he could not do anything with the data unless she/he also obtained the decryption key (which is typically stored on a separate machine).

Typically, when data sets are created for analysis, in most cases, PHI does not need to be part of those data sets. *De-identified* data is data that has been stripped of PHI, though it may still be reidentifiable. In principle, if enough parameters are divulged, one can identify a sufficiently prominent individual if the parameters are sufficiently rare in the general population, and one knows the particulars of the individual from other sources—thus, a cluster of closely related individuals with pancreatic cancer, and no other information other than the ages at death, would identify former President Jimmy Carter's family with high probability.

Anonymized data is data that has close to zero probability of being re-identified. For example, instead of disclosing individual-level de-identified data, one could disclose aggregate numbers only—count, maximum, minimum, mean, and standard deviation. Even here, with a naïve approach, disclosure is possible (eg, if the count of values is disclosed to be three, and the maximum, minimum, and mean are disclosed, one could determine the three individual values).

7.3 PROTECTING DATA

There are three broad approaches, which are employed simultaneously, for protecting data. Protection includes both prevention of unauthorized access as well as protection against loss (due to natural or human-created disasters). The NIST document

SP-800-111 [5] deals specifically with guidelines for encrypting data for storage on end-user devices (including portable devices such as USB keys).

Administrative measures include the following.

- Giving people the data and privileges that they need to do their jobs, but not more (the Principle of Least Privilege).
- Categorizing data (and elements within data) based on importance as well as confidentiality, and having policies in place regarding the appropriate degree of protection for each, as well as the hardware used for data storage. Such policies not only apply to data stored electronically but also paper-based documents. As for most things, all data do not have equal priority: because of cost considerations, high-value or high-confidentiality data should be protected more vigorously.
- Data-use agreements, which hold individuals responsible and liable for both de-identified and patient-identifiable data that is necessary for research, but whose inadvertent disclosure may harm patient privacy.
- Assigning security *responsibilities* (as well as *authority*) to designated individuals.

One problem at Sony's movie division, which was breached in 2014 by intruders supposedly sponsored by North Korea, is that, even after Sony appointed a supposedly organization-wide chief security officer following a 2011 break-in at their PlayStation division, this officer did not have the authority to make changes at the movie division, which insisted on operating in a security silo [6]. Whether granting the requisite authority might have prevented the second attack, however, is questionable.

- Policies on disposal of documents and electronic media, and on sending patient information via e-mail.
- Standard operating procedures for offsite storage of backups, backup frequency, and disaster recovery.
- Policies regarding penalties for accidental or willful disclosure of confidential patient information.
- Enforcement of all of the policies, and periodic tests to evaluate the adequacy of all of the previously mentioned measures.

Physical and technical measures: These include:

- Servers holding confidential data that are housed in secure machine rooms with logged keycard access.
- Judicious employment and management of passwords. Passwords are stored as hashes rather than as plain text. If the password file can be copied, it is vulnerable to "dictionary attack"— an attacker can employ a database containing every word in the dictionary (and sometimes combined with numeric suffixes), and try to hash every single word until a match to an entry in the password file is found. This is why users are dissuaded from using single words as passwords, and are often required to use a combination of upper case, lower case, numbers, and punctuation.

At Sony, one of the passwords employed was simply "password." While seemingly clever, this happens to be a word in the dictionary and so is vulnerable to dictionary attack.

Within a single institution, a user may be required to have access to dozens of resources. This can result in "password fatigue." Further, if each resource requires its own username/password combination, users tend to write them down as a list on paper or electronic media because they are impossible to remember individually; if this list is stolen, then all the effort of creating separate passwords was in vain. *Single sign-on* is a mechanism by which the user logs on to a central system (a "directory") just once—this directory records the access privileges of the user for each separate resource. From that point on, access to other intrainstitutional resources is seamless.

The mechanism employed for single sign-on is a vendor-independent standard called Lightweight Directory Access Protocol (LDAP). The single-administration-point mechanism also has another advantage: if a user's security is compromised, access to all systems can be turned off centrally. However, until such compromise is detected, the intruder has the "keys to the kingdom," and so additional measures ideally need to be employed, as described next.

- *Two-factor authentication* is ideally employed for end-users at login. It uses a combination of something you *have* [a hardware token, a cellphone that the system calls back on (employed by many banks), or a biometric (physical) property of the user, such as a fingerprint, retinal scan, voice pattern] with something you *know* (the password): neither one alone suffices to give you access. Sometimes, an additional mechanism, such as a secret question, is also employed. Note that biometric identification is not as secure as originally believed: thus, fingerprints can be spoofed by being lifted off objects (eg, wine glasses) and reused, while voice patterns lock out users who may be suffering from a bad sore throat or a cold (which changes resonance).

  Two-factor authentication is now employed by many academic centers, notably Yale University, which has long been a favorite target for intruders.
- *Antimalware software and password-protected* screensavers (which prevent access when the authorized user is away) are employed on all workstations. As discussed later, antimalware protects only against known threats.
- *Defense in depth*: This approach assumes that some intrusions will succeed so measures should be instituted that prevent breaches affecting one part of an organization's network from affecting other parts. All accesses should be logged and monitored, with a change in access patterns from unknown external sources setting off alerts.

Training: End-users are by the far the weakest link in security: 80% of all data breaches (inadvertent or deliberate) are due to insiders, due to carelessness (in most cases) or malice (in a few). Consequently, "old-fashioned" methods (ie, stealing encryption keys or passwords through subterfuge, or obtaining them through physical threats or bribes) are more likely to be effective than approaches based purely on computing. This theme is discussed later.

7.4 INSTITUTIONAL PREPAREDNESS

Hicks [7] provides an excellent overview on meeting HIPAA encryption requirements. He emphasizes that HIPAA legislation does not micromanage how data is encrypted: the only *recommendation* (which is not the same as *requirement*) is that unclassified information utilize at least FIPS-validated cryptography (FIPS, Federal Information Processing Standard), while classified information employ NSA-approved cryptography (more on the NSA later).

Interestingly, neither HIPAA nor the more recent Health Information Technology for Economic and Clinical Health (HITECH) act of 2009, which promotes the adoption and meaningful use of healthcare technology, *mandates* encryption of healthcare data, although HITECH provides incentives for encryption. Hick's paper introduces the concept of *safe harbor*. That is, *if* PHI is stored encrypted—which means that it is not decipherable to unauthorized individuals—and there is a guarantee that the encryption key was not stolen (because it was on a separate device) then, in the event of a system breach, the organization does not need to notify all affected patients through mass mailings or notify the Federal Government. Implementing safe harbor requires the organization to track in detail where PHI is stored: even within a database, the PHI-containing fields (which would include data such as choice of secret questions and the users' answers to them) must undergo a layer of encryption in addition to any hardware-based encryption that may be applied to the storage device as a whole.

The NIST document SP-800-66 [8] provides a framework for risk management with respect to HIPAA-required data protection. While fairly repetitious, the basic strategy in dealing with various aspects of security includes the following steps, which are followed in a cyclical fashion: that is, after the final step, one may go back to the first one.

1. *Categorize* data (or systems) based on risk and vulnerability.
2. *Choose* the security controls that should be applied to each resource. Document these choices and the rationale for using them.
3. *Implement* the security controls.
4. *Assess* (test) how well these work and whether they have been implemented correctly.
5. *Go live.*
6. *Monitor* continuing threats and vulnerabilities to the system: you will have to continuously revise your safeguards. When changes to the system have been made, these must be documented. All potential changes must be assessed for impact before they are made.

7.5 HIPAA MATTERS: CALIBRATING THE LEVEL OF PRIVACY TO THE LEVEL OF ACCEPTABLE RISK

An unfortunate side effect of HIPAA is that has given short-sighted IRBs in certain institutions a license to treat researchers as though they were criminals, so that the barriers that are set up interfere with legitimate research. When the Health Insurance Portability and Accountability Act (HIPAA) was first introduced, there was much confusion about the circumstances under which clinical research studies could record PHI. After a wave of idiocy and hysteria, sanity ultimately prevailed as IRBs realized that *patient safety always overrides privacy concerns.* The Joint Commission (for healthcare accreditation/certification) mandates that in clinical care, to minimize the risk of treating the wrong patient, every patient be identified by *at least* two identifiers in addition to full name—typically date of birth and home address. The same applies to clinical studies, such as Phase I–III clinical trials of anticancer agents, where there is the possibility of harm from accidentally entering one patient's data against another, and then making decisions such as dose escalation or drug cessation on that erroneously entered data.

In any case, many of the policies put in place in several institutions have the effect, to paraphrase the Gospel of Matthew, of straining at gnats while allowing camels to be swallowed. Measures that only convey the impression of better security, without actually improving security or thwarting a determined intruder, are described by security expert Bruce Schneier as "Security Theater" [9], which refers to any half-hearted and poorly thought out approach that would work only against extremely unsophisticated malefactors, and is really little more than a placebo for its target audience, the general public.

An example Schneier provides is the practice of the Transportation Security Administration (TSA) of checking boarding passes before letting a potential passenger into the flight waiting areas. The naïve traveler may feel safer seeing such measures in place. However, such passes can easily be faked, complete with bar code, by any reasonably computer-literate person using Adobe Photoshop [10]. More concerning, the boarding passes leak information that can be used by terrorists or malefactors. (I am indebted to Dr Kimberly Dukes for bringing these to my attention.)

- At one point in time, the last digit in the barcode indicated whether the passenger could bypass precheck, or was to be given the usual prescreening. A terrorist could decode the barcode, print a modified barcode with a changed name (to match a fake ID), and the last digit changed to avoid a precheck, and get onto the flight. The TSA's scanners would not pick up the altered information [11].
- The barcode contains last name, flight record locator, and frequent flyer number. For many airlines, the first two, decoded by someone experienced with barcode technology (or by software that does the decoding and can be used by anyone) can provide access to the person's frequent flyer account without entering username and password, whereby a malefactor can view future travel plans (and change them) [12].

Researchers are not the Achilles heel of the organization as far as privacy goes. The fact is that the vast majority of patients are "nobodies": other than the insurance companies (who have that information anyway, and who have repeatedly abused that privilege in the past to deny people insurance), few people care whether Joe Average suffers from heart disease, diabetes, or HIV. And if the target happens to be a public figure or celebrity, there are other, far more economical means of getting at the desired information than hiring a high-priced electronic intruder. The tabloid press, for decades, has used the simple expedient of reaching out to the lowest-paid persons in the hospital staff through financial incentives.

Paul Clayton, former Chair of Biomedical Informatics at Columbia, once stated that in New York City, the price required to breach the privacy of a celebrity admitted to New York-Presbyterian could be as low as $200, paid to the guy who cleaned the celebrity's bedpan.

As discussed toward the end of this chapter, the situation has changed slightly: the nobodies' data is now used by organized crime to defraud a third party, such as the US Federal Government or the credit card agencies, though we all end up paying for such crime indirectly.

7.5.1 Working with narrative text

A practical issue relates to research in *natural language processing*, which refers to methods of extracting information from electronic narrative text. PHI is embedded unavoidably in clinical notes, even if one refers to the last name of the patient in passing. Short of manual methods, it is impossible to remove such PHI with 100% accuracy—a problem is that many last Anglo-Saxon last names are also common words in English or medicine—Black, White, Green, Brown, Sand, Burns, and Blood.

> **There was even a famous neurologist named Russell Brain who, interestingly, wrote a text on diseases of the nervous system. This book, continually revised and still in use, was at one time edited by (Sir) Roger Bannister, who is famous for an athletic record. While practicing medicine as a resident, Bannister became the first person to run a mile in under 4 min.**

Removing every possible phrase ever known to be a last name may mutilate a medical document beyond recognition. The Data Use Agreement is a workable means of balancing the needs of research with privacy, as opposed to not allowing work with narrative data at all. (Otherwise, if an NLP researcher develops a new method of stripping PHI from data, that researcher would not be allowed to use real data to test the method if research on real narrative text data was proscribed.)

7.5.2 Date-shifting: a warning

A similar concern applies to the use of de-identified clinical data marts such as i2b2, which was described earlier. i2b2 is used to identify sizes of sets of patients identified through EHR data (cohorts) who match arbitrary criteria. It is intended to be used by researchers in "self-service" mode (ie, without prior IRB approval). Such sample-size estimations are essential before applying for research grants for clinical studies: eventually, subjects are to be recruited from this cohort.

One of the important criteria here is that the patients in the cohort should be restricted to those seen in the recent past, where the definition of "recent" varies (eg, 1–2 years ago). This is because patients seen in the remote past may not be alive or may have moved to a different geographic location. IRB approval is not necessary because the i2b2 software only discloses aggregate counts of patients and basic demographic characteristics (gender, race, ethnicity, age histogram), also in aggregate form. (The i2b2 software was originally designed by Shawn Murphy's group at Harvard with the input of the Harvard IRB.)

Because dates are considered identifiable data (because specific dates can be tied to specific individuals if we are able to get that information by alternative means), some institutions apply "date shifts" to information in i2b2. Here, every patient in the system has all dates shifted by a random number, from 1–365 days in the past. While date shifting, in theory, makes the system more disclosure-resistant, it drastically reduces the self-service utility of i2b2 because researchers who query i2b2 prior to a grant application using a "time since last visit" criterion are pretty much guaranteed to receive inaccurate (ie, worthless) numbers. They are then obliged to contact the information-technology team who operate i2b2 (and who have access to the actual identifiable data) for help. Therefore, the entire purpose of i2b2, which was to reduce the load on the IT staff by allowing researchers to generate useful numbers by themselves, is defeated.

7.6 A PRIMER ON ELECTRONIC INTRUSION

The battle between electronic attackers and defenders is inherently asymmetrical, and the attackers always have a significant advantage. Marc Goodman succinctly cites the reason: while the software designer is the artist, the attacker is like the critic who only needs to find a single flaw [13].

At the outset, I'll define a few terms.
- *Malware* is simply "malicious software," which compromises the confidentiality, integrity, or availability of data, applications, or operating systems [5].
- *Backdoor* is a weakness, built in or installed after initial compromise of a system, which gives an attacker subsequent access to a system: such access can continue as long as the attacker deems necessary. Attackers can use backdoors, among other things, to cover up their tracks.
- *Trojan* (the name comes from the Trojan horse of Homer's *Iliad*, a subterfuge that won the Trojan War for the Greek allies) is malicious code that a user naïvely installs while believing that it is useful. A *worm* is a program that creates a large number of copies of itself that are transmitted over a network, consuming network resources through a flood of traffic.
- *Virus* is a program that attaches itself to another program and replicates itself. A keystroke logger keeps track of the user's keystrokes (which may include usernames and passwords) and records them, transmitting them periodically to a remote site.
- *Social engineering* refers to methods that rely on compromising security/privacy by leveraging human frailty or gullibility.

These terms are not necessarily mutually exclusive. Thus, trojans may exhibit viral behavior and perform both keystroke logging and backdoor creation: users are induced to install them through social engineering.

The Internet greatly expanded following the High Performance Computing and Communication Act of 1991, also known as "the Gore bill." Tim Berners-Lee's creation of the protocols underlying the world wide web, hypertext transfer protocol (HTTP), and hypertext markup language (HTML), and the 1993 creation of Mosaic, the first graphical Web browser, by Marc Andreesen and Eric Bina (which leveraged these protocols) brought the Internet to the layperson's awareness.

Internet-distributed malware—followed shortly after. Malware and computer crime takes advantage of human frailty, or of design or programming defects in software, or a combination of the two. The former is simpler to exploit and many not even require any significant software skills at all to employ. P.T. Barnum's dictum, "There's a sucker born every minute," has been true since humanoid ancestors descended from the trees in the African Rift Valley.

7.6.1 Social engineering techniques

The Wikipedia article [14] is very informative. For example, in "Phishing attacks," attackers send e-mails claiming to originate from the recipient's workplace or bank to fool them into divulging their credentials. A "spear-phishing" attack, which targets very specific individuals by using content such as personal/company details to increase the appearance of legitimacy, was used successfully for an industrial espionage case targeting the Coca-Cola Company [13].

The naïve disclosure of personal details—family members, spouses or partners, birthdays, recent vacations undertaken—on social networks such as Facebook is often utilized by organized crime, using unskilled labor to glean details that are then employed for either spear phishing or identity theft. Similarly, users may be induced into downloading spyware (which category includes keystroke logging programs or programs that surreptitiously activate any attached video camera) through websites that offer games or pornography. In addition, the tendency of many state agencies to sell personal information on their citizens to any data broker (which could include criminals masquerading as the same) is also to be deplored: it is not too difficult these days to obtain anyone's social security number, often the first step to identity theft.

7.6.2 Notable electronic exploits

The world's first Internet worm was created by Robert Tappan Morris in 1988. Released accidentally, it took down about 10% of ARPANET (the Internet's predecessor), relying on vulnerabilities in the UNIX operating system and on penetrating weak passwords [15]. (Morris avoided jail time because he contacted system administrators after its release and advised them on how to remove it, but he had to pay a hefty fine.) The 2000 ILOVEYOU worm (which also erased image files) relied on flaws in Microsoft Windows and Microsoft Outlook. The worm also relied on the curiosity of users, who opened e-mail messages without being suspicious.

The initial motivation behind malware was simply a form of vandalism, or a desire to show that something could be done (in the manner of Edmund Hillary's supposed motive for climbing Mount Everest, "Because it's there."). It was only a matter of time before the commercial and military possibilities of malware opened up, and malware creation became a full-fledged research effort involving large teams. The most sophisticated malware yet created—Stuxnet, which targeted centrifuges at the uranium-enrichment plant at Natanz, Iran, gradually rendering a fifth of them nonoperational—was supposedly created jointly by the United States and Israeli cyber-warfare groups, possibly working with Siemens staff. (Iran employed Siemens programmable controllers, which they had obtained secretly, to control the centrifuges.)

Just as the most successful parasites are those that cause minimal disturbances in their hosts, the most successful exploits are those that go undetected, Thus, while Stuxnet infected an estimated 60% of all computers in Iran, 8% of all PCs in India, and 1.5% of computers in the United States, it did nothing other than copy itself to removable media (eg, USB key drives) in the overwhelming majority of machines that did not have attached Siemens controllers.

7.6.3 Scope of the problem

Today, operating a desktop computer without antimalware software is unthinkable—though for the most part this protects against known threats rather than previously unknown ones. The problem has spread to mobile technology as well: mobile malware may soon become the dominant form of malware because operating smartphones requires the least amount of electronic literacy.

Experiments by security researchers have shown that the "Internet of Things"—automobiles, medical equipment, industrial robots—are equally, if not more, vulnerable, mainly because of complacency by manufacturers who practice "security through obscurity" [16], a short-sighted approach that relies on not disclosing system internals, rather than systematically forestalling weaknesses during the design process itself. This is already the stuff of Hollywood plot devices, and Vice President Cheney's physicians were concerned enough about such vulnerabilities to disable remote access to Cheney's cardiac pacemaker [17].

The Shodan search engine [18] can discover supervisory control and data acquisition (SCADA) devices—traffic lights, power grids, even nuclear power plants—that are connected to the Internet within a geographical range and can identify vulnerabilities in these systems. Shodan can be used by the bad guys to probe for vulnerable devices that could be taken over, and by the good guys to make sure that their systems do not have any vulnerability. Three German students used Shodan to discover (and obtain read-write access to) about 40,000 databases that ran on the NoSQL engine MongoDB, mentioned in chapter: Core Informatics Technologies: Data Storage, Section 3.5, which used MongoDB's default (low-security) configuration [19]. US security researchers similarly discovered 68,000 exposed medical devices—including infusion systems, MRI/nuclear isotope scanners, and picture archiving systems—across all the hospitals of a large healthcare organization [20].

7.7 STATE OF HEALTHCARE SYSTEMS WITH RESPECT TO INTRUSION RESISTANCE

Healthcare in the United States has unfortunately been profit-driven. The history of the introduction of electronic systems into healthcare-related organizations reflects these priorities. Until about 3 decades or so ago, most electronic systems in healthcare operations were more focused on administrative and financial matters than on patient care: it didn't matter whether the patient lived or died, as long as no bills were outstanding. Mercifully, this has changed somewhat, but in the application of computing technology, healthcare lags way behind most other fields, and the mediocrity of health IT is an open secret: the really smart computing graduates choose to work in startups, or at places like Google or Amazon.

The overall trends seem to apply to health IT security as well, whose weakness is now known to communities of electronic malefactors. (I prefer this term to "hacker," which is actually a term of respect in computing circles, referring to someone who is creative or skilled.) Health IT is now known to be among the weakest links in the security ecosystem, and now healthcare systems are targeted to steal, not healthcare data, but personal information on patients, which are then used in other crimes. If the crooks haven't got at clinical data, it's only because they don't care to.

Two recent articles in the *Washington Post* [21,22] discuss this issue. One of them deals with the Anthem break-in, supposedly organized from China, where intruders made off with the personal information of 80 million customers due to a failure to encrypt the latter's personal data (as recommended by NIST, see earlier). Information disclosed, in addition to PHI such as home address and SSN, also included salaries and places of employment. The Anthem breach was targeted theft—system-administrator credentials were stolen along with unencrypted data stolen from company's servers. Interestingly, though the US Senate decided to review HIPAA security after this breach [23], it is not clear that the US Federal Government penalized Anthem under HIPAA: the maximum penalty under HITECH, in any case, is only $1.5 million [24], which is a rounding error in Anthem's budget. Instead, private parties have filed class-action lawsuits under breach-of-contract and negligence statutes [25].

The only reason why healthcare-data intrusions are not greater is because health information *per se* is far less valuable to the intruder than financial or industrial-secret information.

7.8 ROLE OF THE US GOVERNMENT

The NSA advises NIST on security matters, and its standards are part of the HIPAA rule. After the Edward Snowden disclosures, however, it also became clear that the NSA has been complicit in *weakening* security. Many vulnerabilities in popular operating systems, notably Microsoft Windows, remain unfixed, providing the NSA with backdoors [26]. Further, the security company RSA was paid $10 million by the NSA to make a discredited cryptography system, also with a backdoor vulnerability, the default for their software [27]. Under the PRISM program, the NSA has legally mediated access to Google and Yahoo user accounts, but has also surreptitiously accessed Yahoo and Google data centers worldwide [28].

While the vulnerabilities are initially known only to the NSA, they eventually get discovered by malware creators, who then design tools with graphical user interfaces that can allow unskilled miscreants with little or no understanding of technology (so-called "script kiddies") to use them to intrude [13]. Further, the cyber-espionage units of foreign countries, such as China, have enough resources to discover the very same backdoors and use them against the USA: 21 million personnel records were stolen recently from US Government systems through intrusion believed to be organized from China [29]. The US Internal Revenue Service similarly discovered the extent of identity theft in early 2015 after a flood of thousands of fake early returns claiming refunds, most of which were processed and paid out.

I wouldn't be surprised if the IRS theft is later found to be tied to the Anthem break-in, which was *discovered* only a couple of months earlier. Information stolen for one purpose (eg, identifying Americans of Chinese or Tibetan origin who can then be arm-twisted by using their relatives in China as leverage) is often resold on the underground market to international organized-crime outfits. A faked tax return, in order to be persuasive, should, in addition to incorporating a salary similar to the previous year's, correct home address, and place of employment, ask for a modest amount of money back so that it doesn't arouse suspicion. The Anthem breach could have provided all of the necessary information.

Just like the case of Osama bin Laden, an alumnus of the CIA's 1980s mujahideen anti-Soviet training program in Afghanistan who ultimately turned his skills against his trainers and former sponsors, these are examples of the law of unintended consequences.

The vulnerabilities in existing systems, unfortunately, tend to get fixed only after security researchers or firms discover the same flaws and publicize them. (At one time, the US Government, as well as the corporations concerned, had a "kill the messenger" attitude, where many such individuals were threatened with lawsuits, but mercifully this seems to have changed.) However, disclosure of these vulnerabilities and the NSA's role in maintaining them, as pointed out by Bruce Schneier in his best-selling book *Data and Goliath* [30]—which is essential reading for anyone who values their privacy and civil liberties—has merely made the US software and hardware industry less competitive. International customers increasingly shy away from commercial US technology—in particular cloud computing, which is currently dominated by US providers—following the lead of the Government of Brazil, which abandoned Microsoft Windows and switched to open-source Linux in 2005 [31].

Schneier regards the existing NSA approach as intellectually bankrupt: it works in the short term but fails in the long term. (Incidentally, he regards Snowden as a patriot and hero.) He advocates the *opposite* policy, of government working with vendors to fix their security holes, so that *all* of us become collectively more secure, even if it means that the NSA finds it more difficult to conduct their spying operations.

BIBLIOGRAPHY

[1] D. Kahn, The Code-Breakers: The Story of Secret Writing, MacMillan, London, UK, (1967).

[2] A.C. Doyle, The Adventure of the Dancing Men, The Return of Sherlock Holmes, McClure, Phillips & Co, New York, NY, (1903).

[3] S. Singh, The Code Book: The Science of Secrecy From Ancient Egypt to Quantum Cryptography, Anchor, London, UK, (2000).

[4] Wikipedia. Public-key cryptography. Available from: https://en.wikipedia.org/wiki/Public-key_cryptography, 2015.

[5] K. Scarfone, M. Souppaya, M. Sexton, Guide to Storage Encryption Technologies for End-User Devices, National Institute of Standards and Technology, Gaithersburg, MD, (2007).

[6] J. Gaudiosi, Why Sony didn't learn from its 2011 hack. Fortune Magazine 12/24/2014. Available from: http://fortune.com/2014/12/24/why-sony-didnt-learn-from-its-2011-hack/, 2014.

[7] A. Hicks, Meeting HIPAA encryption requirements. Available from: https://hipaacentral.com/Documents/Perspectives/HIPAA-Encryption-Requirements-Perspective.pdf, 2014.

[8] M. Scholl, K. Stine, J. Hash, P. Bowen, A. Johnson, C. Smith, D.I. Steinberg, An Introductory Resource Guide for Implementing the HIPAA Security Rule, National Institute of Standards and Technology, Gaithersburg, MD, (2008).

[9] B. Schneier, Beyond security theater, Schneier on Security, Available from: https://www.schneier.com/blog/archives/2009/11/beyond_security.html, 2009.

[10] R. Stross, Theater of the absurd at the TSA. The New York Times December 17, 2006. Available from: http://www.nytimes.com/2006/12/17/business/yourmoney/17digi.html, 2006.

[11] C. Doctorow, Aviation vulnerability: scan boarding passes to discover if you're in for deep screening; print new barcodes if you don't like what you find. Available from: http://boingboing.net/2012/10/25/aviation-vulnerability-scan-b.html, 2012.

[12] B. Krebs, What's in a boarding pass barcode? A lot. Krebs on Security 10/15/2015. Available from: http://krebsonsecurity.com/2015/10/whats-in-a-boarding-pass-barcode-a-lot/; 2015.

[13] M. Goodman, Future Crimes, Doubleday, New York, NY, (2015) p. 243.

[14] Wikipedia. Social engineering (computer security). Available from: https://en.wikipedia.org/wiki/Social_engineering_(security); 2015.

[15] C. Stoll, The Cuckoo's Egg (Epilogue), Doubleday, New York, (1989).

[16] Wikipedia. Security through obscurity. Available from: https://en.wikipedia.org/wiki/Security_through_obscurity, 2015.

[17] D. Kloeffler, A. Shaw, Dick Cheney feared assassination via medical device hacking: "I was aware of the danger". Available from: http://abcnews.go.com/US/vice-president-dick-cheney-feared-pacemaker-hacking/story?id=20621434, 2013.

[18] J. Matherly, Shodan: the search engine for the Internet of Things. Available from: https://www.shodan.io/, 2015.

[19] J. Heyens, K. Greshake, K Petryka E, MongoDB databases at risk: several thousand MongoDBs without access control on the Internet. Saarbrücken, Germany: University of Saarland, 2015.

[20] D. Pauli, Thousands of "directly hackable" hospital devices exposed online. The Register (UK) 9/29/2015. Available from: http://www.theregister.co.uk/2015/09/29/thousands_of_directly_hackable_hospital_devices_found_exposed/, 2015.

[21] D. Harwell, E. Nakashima, China suspected in major hacking of health insurer. Washington Post February 5, 2015. Available from: http://www.washingtonpost.com/business/economy/investigators-suspect-china-may-be-responsible-for-hack-of-anthem/2015/02/05/25fbb36e-ad56-11e4-9c91-e9d2f9fde644_story.html.

[22] A. Peterson, 2015 is already the year of the health-care hack—and it's only going to get worse. Washington Post March 20, 2015. Available from: https://www.washingtonpost.com/news/the-switch/wp/2015/03/20/2015-is-already-the-year-of-the-health-care-hack-and-its-only-going-to-get-worse/.

[23] R. Zimlich, Senate to review HIPAA security of medical records in light of Anthem breach. Medical Economics 2/14/2015. Available from: http://medicaleconomics.modernmedicine.com/medical-economics/news/senate-review-hipaa-security-medical-records-light-anthem-breach.

[24] US Department of Health and Human Services. HITECH Act enforcement interim final rule. Available from: http://www.hhs.gov/ocr/privacy/hipaa/administrative/enforcementrule/hitechenforcementifr.html, 2009.

[25] T. Huddleston Jr, Anthem's big data breach is already sparking lawsuits. Fortune Magazine 2/6/2015. Available from: http://fortune.com/2015/02/06/anthems-big-data-breach-is-already-sparking-lawsuits/.

[26] J. Abel, NSA "backdoor" mandates lead to a computer-security freak show: Microsoft Windows OS vulnerable to hackers, thanks to National Security Agency requirements. Consumer Affairs March 6, 2015. Available from: http://www.consumeraffairs.com/news/nsa-backdoor-mandates-lead-to-a-computer-security-freak-show-030615.html, 2015.

[27] J. Menn, Exclusive: NSA infiltrated RSA security more deeply than thought—study. March 31, 2014. Available from: http://www.reuters.com/article/2014/03/31/us-usa-security-nsa-rsa-idUS-BREA2U0TY20140331, 2014.

[28] B. Gellman, A. Soltani, NSA infiltrates links to Yahoo, Google data centers worldwide, Snowden documents say. The Washington Post 10/30/2013. Available from: https://www.washingtonpost.com/world/national-security/nsa-infiltrates-links-to-yahoo-google-data-centers-worldwide-snowden-documents-say/2013/10/30/e51d661e-4166-11e3-8b74-d89d714ca4dd_story.html, 2013.

[29] J. Hirschfield Davis, Hacking of government computers exposed 21.5 million people. The New York Times July 9, 2015. Available from: http://www.nytimes.com/2015/07/10/us/office-of-personnel-management-hackers-got-data-of-millions.html, 2015.

[30] B. Schneier, Data and Goliath: The Hidden Battles to Collect Your Data and Control Your World, WW Norton & Company, New York, NY, (2015).

[31] S. KIngstone, Brazil adopts open-source software March 6, 2015. Available from: http://news.bbc.co.uk/2/hi/business/4602325.stm, 2005.

CHAPTER 8

Mobile Technologies and Clinical Computing

8.1 INTRODUCTION

"Mobile computing" refers to computing applications that arise, or become possible because the devices employed are not bound to a single physical location. As recently as a decade ago, "mobile computing" implied applications involving laptop computers only. At present, the number of smartphones and tablet computers worldwide greatly exceeds the number of laptops. Nearly 64% of Americans in 2015 are estimated to own a smartphone of some kind, and 19% rely on a smartphone to access Internet-based online services [1].

This does not necessarily mean that users have become more computer literate: a 2012 Telefonica (United Kingdom) survey estimated that the four commonest uses of smartphones were, in this order, browsing the Internet, checking social networks and e-mail, playing games, and listening to music. Making telephone calls, the *fifth* commonest activity performed by smartphone users, accounts for only an average of 12 min of the more than 2 h that smartphone users spend on their devices every day [2].

In terms of operating systems employing for smartphones (and their larger-sized relatives, the tablet computers), by the second quarter of 2015, devices using Android (an open-source platform developed by Google) accounted for 83% of all devices, Apple's iOS accounted for 14%, while Microsoft's Windows Phone accounted for 2.6%. However, Apple devices are still the most profitable because the company controls both the hardware and the software, and because the devices sell for a premium.

Mobile computing began with portable computers, with the Osborne 1 (1981) being the world's first mass-produced microcomputer-based portable. The Compaq portable, the first IBM PC-compatible portable, appeared a year later. Initially, primarily because of the weight of the monitor, which used cathode-ray-tube technology, such devices were really "luggable" contraptions that had to be moved around on wheels: the Compaq weighed about 13 kg. The introduction of liquid-crystal display (LCD), and later, light-emitting-diode (LED) technology significant reduced the weight of the display and made true portables possible. Today's largest tablets (as well as the high-end "ultraportable" laptops) weigh about 1.6 lb (0.73 kg) for a 12″ screen: smartphones weigh a fifth as much (140 g). The 2015 iPhones have a screen resolution of 1920 × 1080 pixels with a screen size of 5½″ (14 cm).

Clinical Research Computing. http://dx.doi.org/10.1016/B978-0-12-803130-8.00008-7
159

With the steady shrinking of size, the potential applications of mobile devices have multiplied. Mobile technologies have now become the most widely used electronic devices for *consumption* (ie, viewing) of Internet-based content. The reason for this is, of course, convenience: when a device is light enough to be carried in one hand (or simply worn as eyewear), it is much simpler to integrate content consumption with ordinary activities. Despite the ergonomic drawbacks, discussed later, the freedom to access content from anywhere very often offsets the ergonomic disadvantages to the extent where, even at home, many users will prefer to look up information on their tablet or smartphone rather than walk to the room where their desktop computer is based.

In this chapter, I discuss the role of the following technologies in biomedical research applications.

- *Smartphones and tablets* and the various sensors that interface to them and utilize their processing power. It is interesting to note that Apple's 2011 iPad2, while deliberately limited in features such as local storage, has horsepower comparable to the Cray-2, the world's fastest computer in 1985 [3] (which weighed several tons and was cooled with a special refrigerant). Mobile devices have only become more powerful since then.
- *Laptop computers*, despite the drop in sales since the introduction of smartphones and tablets, these still fill computing niches where the other two devices are not fully satisfactory.
- *Wearable devices*, such as the Apple iWatch, Google Glass, and virtual-reality headsets.

8.2 USES OF MOBILE DEVICES: HISTORICAL AND RECENT

Portable computers began as a way to be able to perform computer-related work anywhere, subject to the limitation that certain resource-hungry software might not run well (or at all) on them because of limitations on memory, processing power, and disk space. As they became progressively more powerful, this limitation gradually disappeared. In many organizations where employees have to move between different physical offices, it is standard practice to issue high-end laptops, and employ docking stations at the offices that connect the laptop to large external monitors and external USB ports.

Touch-screen devices, both portable and nonportable, have long been used in industrial settings as well as in settings such as self-service restaurants and kiosks. These, however, used a single-touch sensing device with gestures limited, for the most part, to pointing or dragging a single finger. The iPhone was the first device that used a multitouch interface, which responded to gestures such as pinching and spreading the finger (used to zoom out or in respectively), as well as discriminating between dragging and swiping (a transient drag where the finger is rapidly withdrawn from the screen).

> **Concurrently, Apple's designers realized that the user interface metaphor for touch-based devices needs to be different compared to desktop devices: the hand is not the same as a mouse. While excellent for hand-held devices, touch screens are unsuitable for continuous use on desktop monitors. Constantly reaching for a screen at arm's length can produce repetitive-injury neck/shoulder strain.**

One of the numerous problems with Microsoft Windows 8, which tried to play catch-up with the iPhone and iPad, was that Microsoft attempted to create a single system that would run on both touch-based devices as well as mouse/trackpad-based devices. Windows 8 was almost unusable on the desktop. As a synonym for "technological disaster," it has been compared with Coca-Cola's infamous 1985 "New Coke" introduction [4].

Handheld devices originally began as highly-scaled down devices that performed such chores as contact management and note taking. Once wireless technology became more widespread, they gradually incorporated mobile-phone and later, e-mail functionality. The current devices, such as those running the Android and iOS operating systems, are essentially computers that also happen to record and play sound, music, pictures and videos, and make phone calls. The quality and ease of use of the image and video recording functions has improved so much as to cause a 40% worldwide drop in annual sales of digital cameras in 2013 [5]. The concern that they may be recorded on cell phone video has already caused changes in police behavior [6]: the courts have long held that such taping is legal as long as it does not interfere with police duties [7].

Wearable devices such as virtual-reality headsets currently have niche uses—notably entertainment and gaming, and potential medical applications such as visualization in 3-D (eg, CT and MRI images). Both VR as well as lower-tech "virtual environments" that use traditional display devices are also being actively explored in educating autistic children to overcome anxiety and phobias [8], acquiring social skills [9], improving their ability to speak in a public setting [10], improving recognition of facial emotions [11], and improving street-crossing skills [12]. (Autism is characterized by significant to profound social-interaction deficits: patients have great difficulty making eye contact and even casual conversation, and find it difficult to recognize emotions in others such as worry, irritation, or confusion.)

Robotics and prostheses are, of course, being extensively applied in treatment of the temporarily or permanently disabled. They are "mobile devices" in the sense that they effectively become part of the patient's body—and help the patient to move autonomously or semiautonomously, and even function productively. An example is the special wheelchair that the scientist Stephen Hawking uses to communicate. This field is a vast and highly specialized area, on which I am not currently qualified to do justice in a few brief paragraphs.

The Google Glass mobile device has the possibility of becoming mainstream. It offers many of the capabilities of smartphones, but with the added advantage of (mostly) hands-free operation—if and when all of the numerous product defects (given later) are worked out. Most medical applications of Glass involve Skype-type applications, with images being transmitted as seen by the caregiver (eg, a surgeon or ophthalmologist), or received from a source such as an online electronic medical record. The "Healthcare Applications" section of the Wikipedia article [13] provides a useful overview.

8.2.1 Using the sensors of mobile devices

The real strength of mobile devices stems from the large number of sensing and other devices that are incorporated, and the fact that most of these are accessible to the software developer to combine in interesting ways. This makes it possible, when combined with the devices' mobility, to create applications that go *beyond* what traditional desktop and laptop devices can accomplish.

Devices on the current generation of smartphones, for example, include several of the capabilities listed next. Note that no device currently has all of the capabilities: some features, like the radio, are disabled at the manufacturing level. Also, other devices may not be programmer-accessible: this may soon change.

- *Telephone* capability (of course) that leverages existing cellular network infrastructure.
- *Clock/alarm-related functionality,* including synchronization with local time, and stopwatch capability.
- *Wi-Fi wireless local-area networking,* as well as *Bluetooth wireless* capabilities. While both are wireless technologies, Bluetooth, a low-power-consumption technology, is intended to allow portable devices to communicate wirelessly, and works over distances of 5–30 m or less (depending on the device). Wi-Fi is a high-power-consumption technology that is intended specifically for wireless networking (ie, Internet access) and can work over 100 m to several kilometers (outdoors—the last by employing high-gain directional antennas). Wi-Fi allows *microlocation* of the user within a restricted space (eg, a hospital ward or department store); the latter has been employed for location-specific messages to users based on their proximity to Wi-Fi beacons placed strategically in different areas of a large store. Similarly, Wi-Fi beacons can be attached to devices that need to be located immediately, but whose physical location may change during use (eg, defibrillators).
- *Audio* capabilities: Monophonic sound (stereo if headphones are used). Audio recording through the built-in microphone as well as (limited) voice recognition.
- A *global positioning system* (GPS), which allows real-time integration with map software, and can also allow the location of a device if it is lost.
- A *proximity sensor,* which disables accidental input due to contact with the face while making a call.
- An *accelerometer,* which measures the linear acceleration of the device.
- A *magnetometer,* which is sensitive to the earth's (and other) magnetic fields. The accelerometer–magnetometer combination allows the device to measure changes in three (X, Y, and Z) axes, where the X- and Y-axes refer to the short and long dimensions of the phone, and Z to an axis perpendicular to the display. This allows a means of detecting the device orientation, and correcting for tilt during image or video capture. The accelerometer can also be used as a pedometer to measure the motion of the user.
- A *vibrational gyroscope,* which measures rotation. This is used for a variety of tasks such as image stabilization during image/video capture and detection of movement by the user.
- A *barometer,* which measures ambient air pressure. It is useful for detection of weather changes (pressure drops during rain and storms) and also functions as an *altimeter* to measure elevation.
- A built-in *FM radio chip* (on the same board that has the Bluetooth and Wi-Fi chips). In the iPhone (though not in some other mobile devices, such as specific Nokia phones), it is currently disabled.

- A *light sensor,* which measures brightness in the vicinity, and is used by the camera to compensate for variations in brightness (eg, automatically activating the flash during image capture). (The brightness of the display is also adjusted based on the ambient light.)
- One or more universal serial bus (*USB*) slots that allows devices such as electrocardiogram (EKG) or electroencephalogram (EEG) leads, glucometers, and pulse oximeters (which detect oxygen levels through the skin of the fingertip or the earlobe) to provide input.

8.3 APPLICATIONS IN BIOMEDICAL RESEARCH

Mobile devices have opened up a vast number of possibilities for novel applications in research and healthcare. The term "mHealth" ("m" for mobile) has consequently been coined, and the online-only, open-access *Journal of Mobile Technology in Medicine* (www.journalmtm.com) is devoted exclusively to such themes. Its content is worth browsing.

I discuss applications of mobile technologies in biomedical research under two categories: obvious, traditional applications that are a direct result of mobility and Internet connectivity, and novel applications that are based on the combination of sensor technologies and processing power.

8.3.1 Traditional applications

Mobile technologies have greatly improved the quality of research data collected in the field because of the increasing availability of cell phone signals and wireless Internet access, with speeds that approach those of dedicated cable.

> **The United States lags considerably behind many other countries both with respect to comprehensive geographic coverage as well as signal transmission speeds. In South Korea, for example, average Internet speeds (21 Mb/s) are twice as fast as the United States' 10 Mb/s, and the country is rolling out 1 Gb/s (1000 Mb/s) networks in urban areas, priced at $20/month.**

Smartphone or tablet-based telemedicine, using software such as secure equivalents of Skype™, are also being employed to deliver care for a geographically distributed patient population, especially in rural areas or areas that are underserved with specialist providers. Here, of course, mobile devices offer the edge over dedicated workstations because they can operate from the patient's own home.

Primary paper-based data collection, which offers no interactive validation capabilities at all, is therefore gradually being eliminated in favor of data collection on mobile devices that act as clients to a remote server-based application. I consider two categories of data-entry applications here, patient-entered data and physician-entered data.

8.3.1.1 Patient-entered data

Patients are often asked to respond to questionnaires, such as surveys. Also, many self-rating scales are employed in psychiatry, such as the CES-D scale [14], developed to assess the severity of depression (which often occurs secondarily after disfiguring surgery or chronic illness). It can work best for everybody if a patient is allowed to enter such data from the comfort of one's home, rather than having to travel a provider's office or hospital, or schedule a phone-based interview.

There are some practical considerations involved here. A basic requirement for self-entered data, apart from a reliable Wi-Fi or cellular connection (which does not exist in all geographical areas) is that the patient be literate. Previous computer fluency, interestingly, is not so critical. Smartphones and tablets have been designed to be readily learned, which is the main reason why people who have never used a computer can often become comfortable with such devices after only about 15–30 min of training. (This ease of use is also the main reason why many users of smartphones are not even aware that they are using computers.)

There are some downsides as well. Because of the inferior ergonomics on smartphones, due to screen size and keyboard limitations (discussed shortly), data entry works best when most or all of the data elements to be captured are based on choices or numbers. (With laptops, of course, this is less of an issue.) A related issue is that in most data-entry scenarios, it is highly desirable that the user be able to review what was entered and make corrections if necessary. With smartphones, the user spends much of the time navigating from one very small screen to another; the use of tablets (whose screen resolution now compares with laptops) is preferable by far.

The fact that patients *can* enter data does not mean that one must make them do so indiscriminately. First, not all patients are fluent with real or virtual keyboards, though with children being increasingly exposed to computers, at least in the United States, this proportion is steadily increasing. Second, if patients are forced to enter vast amounts of data manually, the researcher will quickly wear out her/his welcome, and the patient drop-out rate will be very high.

8.3.1.2 Provider-entered data

Many of the considerations for patient-entered data apply to provider-entered data as well. While physicians may find it more convenient to be able to have information at their fingertips with mobile devices, such devices *do not* reduce their data-entry burden in any way. (As I explain shortly, data-entry ergonomics can be significantly worse.) Many physicians complain that, thanks to the emphasis on documentation, they are now being turned into data-entry clerks.

The more enlightened hospitals have realized that patient satisfaction (and revenue) is maximized if physicians spend most of their time interacting with patients rather than with computers, and therefore employ *medical scribes*—stenographic assistants trained to operate the EHR interface, who have also learned medical terminology—who use laptops or desktops to capture data and orders as the physician dictates. Tablets with keyboard attachments are being increasingly employed for this purpose as well.

8.3.2 Sensor-based applications

Connectivity to external sensors is where the real potential of mobile devices lies in medical applications. Unlike manually entered data, there are no limits on data from a variety of sensors attached to the patient that could be streamed over the Internet (over secure protocols, of course). This is particularly important in care of patients with limited mobility, such as the elderly or disabled.

- Mobile EKG recording and basic EKG diagnosis programs have already been approved by the FDA and are being routinely employed.
- Applications that sense electroencephalographic rhythm have been employed in biofeedback: meditation states produce increased low-frequency (alpha and theta wave) activity [15].
- Sensors that detect the temperature of the hands (increased blood supply to the hands correlates with warming and increased mental relaxation) are useful for lowering the frequency of migraines. Similarly, lowering tension in the muscles of the shoulders, neck, forehead, and jaw helps to relieve tension-type headaches [16]. While biofeedback itself was devised in the 1960s, the availability of applications running on mobile devices makes it possible for patients to readily apply these methods at home in a highly affordable way.
- Sensors measuring galvanic skin response (GSR), where the skin's electrical conductance increases with anxiety due to microsweating, are components of "lie-detector" machines. While such machines have been discredited and are not accepted as evidence (people can learn to control their response), applications that measure it have been sold to provide feedback that helps people control their somatic reactions to stress by learning how to relax actively.
- The Sway Balance Mobile Application [17] is an FDA-cleared balance testing system which uses the built-in triaxial accelerometers of a mobile electronic device to objectively assess postural movement (which is impaired immediately after a concussion).
- Wearable devices such as the Nike Fuelband and Fitbit are useful to motivate people who wish to get more exercise. Note that these devices are not particularly accurate: Nike now has an iPhone application that eliminates the use of the Fuelband altogether, measuring activity directly with the iPhone's sensors [9].
- Through the USB slot, mobile devices can be linked to a USB hub—a device that allows multiple USB devices to be plugged in. Such hubs are potentially useful in setting up mobile ICUs (eg, during disasters or on battlefronts) where multiple sensors connected to a patient must be fed to a computing device.

Steinhubl et al. [18] point out a practical issue related to sensors (and mobile medical applications in general): thousands of applications exist, but only about 100 or so have been reviewed by the FDA. Many of the others are based on quasi- or pseudoscience, and it is very much a matter of "buyer beware."

8.4 LIMITATIONS OF MOBILE DEVICES

With the ubiquity of smartphones, social critics have bemoaned the phenomenon of "Internet addiction" (especially to social media or computer games) and "glassholes," where people spend far more time interacting with their devices than with the world and people around them. There have also been concerns about radiation exposure to cell phones, but the jury is still out [19]. These issues, along with related themes of privacy and acceptable use of recorded content, have been explored in depth elsewhere. Therefore, in this section, I'll focus on ergonomic limitations only.

At one time, the rapid rise of mobile-device use had certain computer industry pundits forecasting the death of the desktop computer. (Such articles still keep popping up; see [20].) However, such claims ignore one critical fact. The limiting factors, as far as mobile devices go, are not the CPU or RAM capacities of mobile devices—which are fully adequate—but the limits on human physiology and the peripheral devices that have been invented to accommodate it: keyboard, visual displays, mice, and other accessories.

Mobile devices, at present, fall considerably short as tools for electronic content *creation*, especially when working for long periods of time. The limitations are now discussed under individual categories.

8.4.1 Display issues

It is well known that one of the major factors that influence productivity, in white-collar folks who have to use computers for a living, is the size of the display [21]. Two or even three large monitors per desktop device have become standard: the more you can see at one time (if you have the desk space), the more the number of applications and documents that you can work with simultaneously. The limits on laptop real estate have accounted for the popularity of docking stations, and similar limitations on the display of the iPhone motivated the introduction of the iPad. As for the Internet experience offered on the 1st generation (non-Apple) smartphones such as the original Samsung Gusto, which has a 1.5″ × 1″ screen, it is astonishing that it is offered at all: the experience is so miserable as to more or less constitute a cure for Internet addiction.

While mobile displays are improving in resolution continuously, a significant proportion of people above 50 find it difficult to read text at the 10-in. distance that is typical for smartphone devices without corrective lenses: they find it preferable to use large monitors where the magnification is enlarged.

Changing magnification is something that most software, but not all, lets you do effortlessly. One of the problems with the now-deceased Google Health website is that much of the content, created by 20-something engineers with excellent corrected eyesight, used fixed-point, nonexpansible 6–8 point font sizes, even though it was intended for a user base with a significant proportion of elderly patients. *This design was employed at a time when web technologies such as Cascading Style Sheets let the developer specify relative sizes effortlessly.*

One of the problems with Google has always been that, while they are highly innovative, they tend to rush products out the door with minimal usability testing with a variety of users. As discussed shortly, many usability problems plagued Google Glass, and the decision to withdraw it from the market (until the technology was improved) was the right one.

8.4.2 Keyboard input issues

For content creation, the one-or two-finger typing required by smartphones and tablets can be a drag on touch typists (a category that increasingly includes the vast number of people who grew up using computers in school). Despite a statement by the late Steve Jobs that the iPad would always have a virtual rather than a physical keyboard, Apple itself seems to have realized that virtual keyboards cannot be used productively for extended periods of time: it now manufactures keyboards that connect wirelessly to the iPad. Third-party keyboard accessories have been available for the iPad long before Apple decided to change its ways.

> **For pictographic languages such as Chinese and Japanese, which include thousands of symbols, a stylus is actually preferable to keyboarding as a means of text input: most people's fingers are a bit too fat to edit text unless the very largest font size is used.**

Limitations of alternative pointing devices—trackpads: Some ergonomic issues have existed since the days when laptops were the dominant mobile devices. Thus, trackpads, originally devised as alternatives to the mouse for use in circumstances of inadequate table-top space, still cause the mouse to move unpredictably in response to accidental grazing by the thumbs of touch typists, or after vibrations caused by an occasional vigorous keystroke. After infuriating a generation of early laptop adopters, most manufacturers learned to make laptop keyboards that include a keypress combination, or a software or hardware switch that disables the trackpad and lets the user employ a mouse exclusively.

8.4.3 Limitations of voice input

The less the likelihood of being able to employ either a real or a virtual keyboard (as for wearable devices), the more reliable voice input must be for the device to be usable. Despite the numerous advertisements with Hollywood actors that have touted this technology, reliability is one problem that has not yet been fully solved: a 2013 survey indicated that 46% of US buyers of the iPhone felt that Apple had "oversold Siri" [22]. Speech recognition on mobile devices works only for very simple commands and is not accurate enough for content creation. The reasons for this have to do with the differences in the ways speech recognition is implemented by dedicated, desktop-based voice recognition versus how it is implemented on mobile devices.

Dedicated speech-recognition software is *speaker-dependent*: it lets you train the software to recognize the way you pronounce words. You employ an initial training period where you recite one or more specific passages. In addition, the more you use it, the more the software adapts to you and the more accurate it becomes because it remembers your corrections. As such, even if you have a heavy foreign or regional accent, the accuracy of the system gradually improves.

Further, vendors such as Nuance (which has also developed the Siri technology used by Apple) also sell medical and legal editions of their software. Interestingly, the special editions are more accurate than the all-purpose version because there are very few phrases that sound like "penicillin," "angina pectoris," or "habeas corpus," even when spoken with markedly varying accents. It is the monosyllabic words that contain homophones—differently spelled words that sound alike, such as "to," "too," and "two" or "sole" versus "soul"—that are hardest to discriminate. Speech recognition handles such words probabilistically, by recognizing the longer words and the corresponding parts of speech in the *vicinity* of the problem word and then guessing what the problem word might be. Thus, to use an over-simplified example, "two" (an adjective) would often be followed by a plural noun, while "too" (an adverb) would be followed by an adjective, such as "cold," "far," or "long."

Smartphone-based speech recognition employs some local processing, but most words appear to be sent remotely to a cloud-based server, which employs "speaker-independent" technology. The service identifies/guesses the words to which your voice signals correspond, based on its statistical model of what individual words tend to sound like. It is not known how much of the natural known variation in pronunciation the statistical model can accommodate. (eg, the English and New Englanders tend to drop their terminal "r's," and Texans are believed to pronounce "all" and "oil" almost indistinguishably.) It is possible that the cloud servers may employ locale-specific statistical models, but this is not known.

Therefore, if your accent is similar to the predominant accent in the locale that has been employed for the statistical model, the server's guess will be accurate. If your accent differs markedly, it will fail. (Siri reportedly has problems with Scottish, Southern, or Boston accents.) More important, repeated use of the device will *not* make the recognition better: the server receives input from hundreds of thousands of speakers every day, as opposed to dedicated software, which is listening only to you. If accuracy improves over time, it does so modestly, and only to the extent that *you* start to change the way you speak.

8.4.4 Specific concerns for wearable devices

Some of the problems experienced by Google Glass users are nicely described in an article by Rohrs [23]. These include extremely poor battery life, less-than-stellar speech recognition, inadequate sound (sound is transmitted to the inner ear via the mastoid process of the skull, which is the bony projection behind the ear, using bone conduction—which doesn't work well is there is ambient noise), reduced screen visibility in bright sunlight, and a clumsy web-browsing experience. The question Rohrs raises legitimately is what Glass offers that your existing smartphone doesn't do already—other than hands-free operation, which is occasionally an advantage but mostly a curse, because you are forced to use voice navigation almost exclusively. (The touchpad located next to the user's right temple lacks the discrimination of a smartphone or tablet.)

One of the challenges of using virtual-reality headsets is that sustained use can bring on nausea or headache, with symptoms similar to motion sickness: Wikipedia actually has a topic on "virtual reality sickness" [24]. This appears to be due to a slight delay between the movement of the eyes or head and the change in the image that results from the hardware loading what is intended to be displayed. As computing processing power improves, this limitation may gradually disappear.

8.4.5 Security issues

A major concern with mobile devices, particularly smartphones and tablets, is security. Security is now even more of a concern than for desktop machines, for several reasons.

- Those who tend to use smartphones exclusively (for e-mail, web browsing, social networking, and games) tend to be much less computer savvy than desktop-machine users. Further, heavy social media users are far more trusting of software vendors and social media: they often divulge the most intimate details of their lives on Facebook or Twitter—including their date of birth and pictures of their family members, which then invite identity theft.
- Smartphone vendors were historically very naïve, ignoring the possibility that malefactors would ultimately target their platforms. For a long time, Google's Play Store (for downloading Android applications) had a reputation as the "Wild West" until Google decided to scan for and remove malware. Apple's weak standard operating procedures for securing iCloud accounts (which are used to backup data stored on iPhones and iPads if the customer chooses) were exploited in late 2012 to take over the online accounts of Matt Honan, a technology reporter for Wired Magazine, and erase 8 years' worth of e-mail and laptop data, as described by Honan himself, who also admits to failing to take some basic security precautions [25]. (The person who hacked Honan described the technique employed to him.)

    As recently as Sep. 2015, Apple discovered that its App Store had been infected with hundreds of malware applications inadvertently created by developers in China who downloaded a fake version of Apple's Xcode mobile-application-development platform (even though Apple gives Xcode away freely). The fake platform, called XcodeGhost, introduced malware that would steal data from users (most of whom were also based in China) [26]. Xcodehost was designed cleverly enough to fool Apple's quality-testing team, which vets applications before allowing them to be deployed. Interestingly, the Snowden disclosures had revealed that the CIA had devised this method of compromising iPhones as early as 2012 [27].

- Android phones have been particularly vulnerable to hacking. This is because individual phone vendors customize the base Android software developed by Google. The problem arises when Google releases new versions of Android that fix security bugs: vendors are often slow to revise their software accordingly, and many make it difficult for users to upgrade their operating system software. (By contrast, when Apple releases a new version of iOS, upgrading is painless.) As a result, versions of Android on individual user's phones may be years out of date, inviting opportunistic malware to take over an unpatched device.
- Techniques of eavesdropping on cell phones are well known. Mobile devices that masquerade as cell phone towers can be driven around in trucks to vacuum all cell phone conversations within a particular radius. Also, features such as Find My Phone (on iPhones), originally intended to locate (and disable) stolen phones, can also be turned against the owner, as can cell phone locator services provided by cell phone carriers or even third-party applications.

Numerous eavesdropping applications exist that can be installed on cell phones, and which can even operate the camera remotely.

The ease of cell phone geolocation is so well known that the first action of staff at battered women's shelters is to take a cell phone from a victim and remove the battery and SIM card, because their abuser may be using the cell phone signals to stalk them. The victims are also warned not to use Facebook, which allows pinpointing of physical location [28].

- All mobile devices, from laptops downward, are vulnerable to eavesdropping if sending signals from Wi-Fi hotspots that are unsecured and do not require password access (eg, at airports or restaurants). The Bluetooth wireless protocol for wireless networking is also particular vulnerable to eavesdropping by intruders within the premises. Even ostensibly encrypted Wi-Fi protocols such as WEP and WPA are vulnerable (though WPA2, the latest protocol, is considered reasonably strong for the moment).

  One should not try to perform operations like online banking from public places unless first connecting to a secure intermediary (if available) using a virtual private network (VPN), which is intended to work despite eavesdropping (like https). Similar concerns apply to patient data sent over Wi-Fi. (EHR technology delivered over the web routinely employs https.) Even here, the most popular VPN solution, the Point-to-Point Tunneling Protocol (PPTP), is now considered insecure.

BIBLIOGRAPHY

[1] Pew Research Center. U.S. smartphone use in 2015. Available from: http://www.pewinternet.org/2015/04/01/us-smartphone-use-in-2015/, 2015.

[2] Telefonica UK Ltd. Making calls has become fifth most frequent use for a smartphone for newly-networked generation of users. Available from: http://news.o2.co.uk/?press-release=making-calls-has-become-fifth-most-frequent-use-for-a-smartphone-for-newly-networked-generation-of-users, 2012.

[3] J. Markoff, The iPad in your hand: as fast as a supercomputer of yore. The New York Times 5/09/2011. Available from: http://bits.blogs.nytimes.com/2011/05/09/the-ipad-in-your-hand-as-fast-as-a-supercomputer-of-yore/, 2011.

[4] R.X. Cringely, Windows 8 as New Coke? That's an insult to New Coke. InfoWorld 05/13/2013. Available from: http://www.infoworld.com/article/2614548/cringely/windows-8-as-new-coke--that-s-an-insult-to-new-coke.html, 2013.

[5] T. Barribeau, More doom and gloom: camera sales dropped badly in 2013, 2014 off to poor start. Available from: http://www.imaging-resource.com/news/2014/03/06/more-doom-and-gloom-camera-sales-dropped-badly-in-2013-2014-off-to-bad-star, 2014.

[6] J. Serna, With smartphones everywhere, police on notice they may be caught on camera. Los Angeles Time 4/22/2015. Available from: http://touch.latimes.com/#section/-1/article/p2p-83349503/, 2015.

[7] J. Serna, P. McGeevey, New push to protect people who videotape police officers. Los Angeles Time 4/22/2015. Available from: http://touch.latimes.com/#section/-1/article/p2p-83358750/, 2015.

[8] M. Maskey, J. Lowry, J. Rodgers, H. McConachie, J.R. Parr, Reducing specific phobia/fear in young people with autism spectrum disorders (ASDs) through a virtual reality environment intervention, PLoS One 9 (7) (2014) e100374.

[9] P. Mitchell, S. Parsons, A. Leonard, Using virtual environments for teaching social understanding to 6 adolescents with autistic spectrum disorders, J. Autism Dev. Disord. 37 (3) (2007) 589–600.

[10] W. Jarrold, P. Mundy, M. Gwaltney, J. Bailenson, N. Hatt, N. McIntyre, K. Kim, et al. Social attention in a virtual public speaking task in higher functioning children with autism, Autism Res. 6 (5) (2013) 393–410.

[11] E. Bekele, J. Crittendon, Z. Zheng, A. Swanson, A. Weitlauf, Z. Warren, N. Sarkar, Assessing the utility of a virtual environment for enhancing facial affect recognition in adolescents with autism, J. Autism Dev. Disord. 44 (7) (2014) 1641–1650.

[12] M. Saiano, L. Pellegrino, M. Casadio, S. Summa, E. Garbarino, V. Rossi, D. Dall'Agata, et al. Natural interfaces and virtual environments for the acquisition of street crossing and path following skills in adults with autism spectrum disorders: a feasibility study, J. Neuroeng. Rehabil. 12 (2015) 17.

[13] Wikipedia. Google Glass: healthcare applications. Available from: https://en.wikipedia.org/wiki/Google_Glass#Healthcare_applications, 2015.

[14] L. Raloff, The CES-D Scale: a self-report depression scale for research in the general population, Appl. Psychol. Meas. 1 (3) (1977) 385–401.

[15] J. Lagopoulos, J. Xu, I. Rasmussen, A. Vik, G.S. Malhi, C.F. Eliassen, I.E. Arntsen, et al. Increased theta and alpha EEG activity during nondirective meditation, J. Altern. Complement. Med. 15 (11) (2009) 1187–1192.

[16] G. Lipchik, Biofeedback and relaxation training for headaches. Available from: http://www.achenet.org/resources/biofeedback_and_relaxation_training_for_headaches/, 2015.

[17] Sway Medical Corp. Sway—balance/reaction time/concussion management. Available from: http://swaymedical.com/system/balance, 2015.

[18] S.R. Steinhubl, E.D. Muse, E.J. Topol, Can mobile health technologies transform health care?, JAMA 310 (22) (2013) 2395–2396.

[19] R. Bilton, The health concerns in wearable tech. The New York Times 3/18/2015. Available from: http://www.nytimes.com/2015/03/19/style/could-wearable-computers-be-as-harmful-as-cigarettes.html, 2015.

[20] J. Valcarcel. In less than two years, the smartphone could be your only computer. Wired 2/10/2015. Available from: http://www.wired.com/2015/02/smartphone-only-computer/, 2015.

[21] R. Ball, Three Ways Larger Monitors Can Improve Productivity: Upgrading the Human Component for Increases in Human Performance. Graziadio Business Review. Pepperdine University 2010,13(1).

[22] Intelligent Voice. New Poll: Apple has "Oversold Siri", say 46% of Americans. Intelligent Voice, 2013.

[23] JK. Rohrs, Confessions of a Google Glass explorer: my likes & dislikes (part 4 of 5). Available from: ExactTarget.com, 2013.

[24] Wikipedia. Virtual reality sickness. Available from: https://en.wikipedia.org/wiki/Virtual_reality_sickness, 2015.

[25] M. Honan, How Apple and Amazon security flaws led to my epic hacking. Wired Magazine 08/06/2012. Available from: http://www.wired.com/2012/08/apple-amazon-mat-honan-hacking/, 2012.

[26] D. Lee, Apple's App Store infected with XcodeGhost malware in China. Available from: http://www.bbc.com/news/technology-34311203, 2015.

[27] M. Lee, Apple's App Store got infected with the same type of malware the CIA developed. Available from: https://theintercept.com/2015/09/22/apples-app-store-infected-type-malware-cia-developed/, 2015.

[28] A. Shahani, Smartphones are used to stalk, control domestic abuse victims. Available from: http://www.npr.org/sections/alltechconsidered/2014/09/15/346149979/smartphones-are-used-to-stalk-control-domestic-abuse-victims, 2014.

CHAPTER 9

Clinical Data Repositories: Warehouses, Registries, and the Use of Standards

9.1 INTRODUCTION

Clinical data repositories are databases intended to facilitate arbitrary querying of the data and analyses for reporting and research. They are *secondary* databases, that is, they receive data that has been originally input into other sources. Repositories can be sub-classified by function into the following categories: ODSs, data warehouses/data marts, and clinical registries. I discuss each category later.

Repositories are populated either electronically by a process called extraction–transformation–load (ETL), which is explained shortly, or with a significant manual component (ie, abstraction of the electronic record). Manual abstraction continues to be employed for registries, as discussed shortly.

9.2 OPERATIONAL DATA STORE

I use this term in the singular form because usually only a single copy exists (or should exist, ideally). This is simply a database where data from multiple separate sources has been brought together into a single physical location. The ODS is where the ETL process occurs.

It is necessary to segregate the ODS into physical/logical compartments based on the "refinement" of the data.

1. The *raw-data* tables closely resemble the original sources from where they were copied: however, they will typically contain a subset of the original data elements. (Only those elements deemed necessary for analysis/reporting are selected.) These tables will undergo a process of "cleansing" (ie, eliminating errors or inconsistent values). (Ideally, errors should be fixed at the source, but in the case of EHRs, that may not always be possible. This is in part because every record, inconsistent or not, is electronically signed; also, the errors/inconsistencies are discovered years after they were created, and there is no possibility of discovering what the correct value should have been.)

2. Inconsistent values are either transformed to be consistent (eg, with a set of permissible values for that field) or, if they cannot be fixed, will have values (or even rows from the table) eliminated. The resulting tables are said to be *cleansed*. These tables may also have extra columns added that contain controlled-vocabulary codes that may not have been part of the original data. (Thus, medications and laboratory test may be mapped to National Library of Medicine's RxNorm and the Logical Observations Identifiers, Names, and Codes (LOINC) vocabularies respectively.)

Clinical Research Computing. http://dx.doi.org/10.1016/B978-0-12-803130-8.00009-9

The cleansed tables are usually in "third-normal" relational form. (For those unfamiliar with this term, third-normal-form tables are organized so that related information is stored across multiple tables with minimal redundancy: consult books on relational database design such as *Information Modeling and Relational Databases: From Conceptual Analysis to Logical Design* [1] or, if you're in a hurry, my lecture notes on the subject of database normalization [2]. Third-normal-form design is preferred for transactional systems.)

3. *Restructured* tables will typically combine fields from multiple tables, and the structure of the data may be changed considerably from the original. One common transformation (which is required for the i2b2 data mart [3] is to convert columnar data into a row structure. Restructured data will be transferred to the warehouse or mart, discussed in the next section.

The process of creating "raw-data" tables from the original sources is called *extraction*, while cleansing, coding, and restructuring are steps involved in *transformation*. Note that these three steps are not necessarily performed in that order: it depends on the data. For example, sometimes it may be more convenient to cleanse after restructuring.

9.3 DATA WAREHOUSES AND DATA MARTS

These databases receive data from the ODS—specifically, the final "restructured" tables—and store it in a structure that is highly optimized for rapid querying. The creation of a warehouse or mart constitutes the final "load" step of ETL. Optimization involves the addition of indexes, and sometimes precomputation of aggregates (eg, counts, sums), so that instead of calculating these statistics over and over again with each query, they are just computed once, and then looked up when required subsequently.

The difference between a data warehouse and a mart is primarily one of scope. A mart contains information only about a specific topic (eg, everything about patients) while a warehouse ideally contains information about the entire organization. Sometimes, the choice of technology may differ: for example, marts often use "multidimensional" database technology that supports extensive precomputation of aggregates, while warehouses almost always use high-end relational technology. (NoSQL systems, however, are starting to be employed for specific types of data, at both the ODS and warehouse/mart level).

Relatively few institutions have warehouses of organization-wide scope because of the challenges involved in such an extensive effort. (One "big-bang" warehouse effort I know of failed miserably and cost the chief information officer his job.) Most organizations work incrementally or in parallel, maintaining multiple data marts, each on a different theme. This is not the ideal situation—different efforts can end up coding the same data element in different ways, so that the two systems will find it difficult (or impossible) to interoperate with each other in the future if they need to. However, the multiple-mart approach is a means of lowering overall risk, and is also prevalent simply because different groups of individuals have different focus (and expertise), and do not necessarily operate as one giant team. I return to this issue of division of responsibility later because it can be a stumbling block in EHR-related research efforts.

For example, a clinical research informatics group may implement an i2b2 data mart, but does not particularly care about inventory and purchasing. Also, while they may collaborate with hospital IT, the groups don't necessarily report to the same administrative authority. (In many institutions, the university, which typically pays the salaries of the clinical informatics team, and the hospital, which pays their own IT staff, are two separate administrative entities. Only a few institutions—Vanderbilt University being among these—have the clinical informatics group also running IT.)

In the succeeding section, I use the term "warehousing" (for conciseness) to refer to the process of implementing a warehouse or mart. Both warehouses and data marts are read-only systems. Their contents are typically refreshed in bulk on a regular basis—once every few hours to once a week or even longer, depending on the criticality of real-time reporting needs—and never changed until the next refresh.

9.3.1 Warehouses and marts designs

One common misconception is that all complex queries are run only against the warehouse or data marts. While these systems are optimized for the commonest form of reports, or for special purposes—thus, i2b2 is designed primarily to allow end-users to perform sample-size estimates for patients matching arbitrary clinical criteria in what-if mode—they are not necessarily ideal for all analyses. This is because when you optimize a database for one purpose, you often deoptimize it for another purpose.

Data marts, for example, commonly employ a "star schema," where a central "fact table" is surrounded by "dimension tables" that serve the purposes of lookup, and which surround the fact table like the spokes of a star. Both sets of tables are often denormalized (ie, deliberately redundant), combining information from multiple source tables to improve performance, by eliminating the need to look up information in multiple places each time.

In the i2b2 clinical data mart, for example, a central "observations fact table" contains details of individual clinical observations. Information such as the patient to whom the observation applies, the clinical concept involved, the visit where the observation was recorded, and the provider who documented the observation are all recorded as IDs. The details of the patient, provider, visit, and concept are each recorded in a dimension table (Fig. 9.1). (For brevity, all the details of the tables have not been shown, and the column names are not necessarily the same as in the original tables. See the documentation at www.i2b2.org for details.)

However, this process of homogenizing the data often makes it hard to see the connections between disparate items, and having something closer to the original data makes many analyses easier. Bill Inmon, in his book *Data Warehouse Performance* [4], identifies two kinds of users who perform queries. "Farmers" are end-users who want periodic reports containing exactly what they need. "Explorers" are data analysts who need to perform numerous one-off queries: when starting off, they have a rough idea of where they want to go, but will typically end up retrieving what they need through trial and

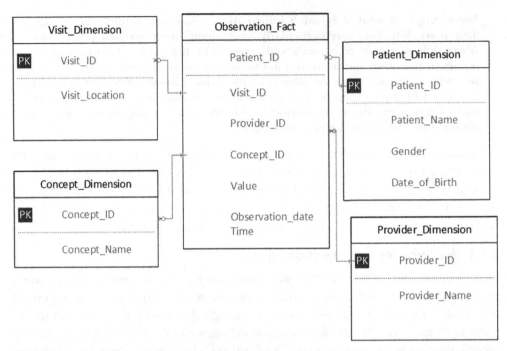

Figure 9.1 *A star schema illustrated with the i2b2 data mart.* For brevity, only some columns have been shown.

error. While "farmers" are best served by star schemas, analysts/explorers are best served by third–normal-form designs, such as exist in the ODS.

9.3.2 The mechanics of data warehousing and ETL

The most useful books I've come across on this subject are the Kimball group's *The Data Warehouse Lifecycle Toolkit* [5] and *The Data Warehouse ETL Toolkit* [6]. All these sources represent hard-earned experience rather than mere theory: read Marc Demarest on "The Politics of Data Warehousing,"[7] for example. I refer you to those resources for details and will limit myself to warehousing issues related to clinical data.

9.3.2.1 Incremental extraction versus full extraction

Ideally, the extraction of data from source systems should be *incremental*: that is, only the changed/added/deleted data from production transactional systems is transferred to the ODS, rather than all of it. This can result in significant efficiency when the source data is several terabytes in size, but the daily changes are of the order of megabytes. However, many EHRs (especially those that employ nonrelational technology) do not support incremental updates.

For incremental updates to work, every record in every table must have three tracking fields: date/time of creation, date/time of last change, and date/time of deletion. Records are never physically deleted but only flagged as such, being moved to archival storage if necessary. Some RBDMSs, such as SQL Server 2016, support "temporal tables": designating a table as temporal will add these fields automatically, and the RDBMS updates these fields invisibly, maintaining a history of all changes, without the developer having to write any special code. However, even if the RDBMS does not support temporal tables, it is not too difficult to write software that implements the same. The problem is that many EHR vendors omit to do this so that, for example, it is impossible to get a history of address changes—which may be important in research related to the environment in which you want to find all people who lived in a particular geographical location at one time, even if they have since relocated.

9.3.2.2 *Physical integration versus virtual integration*

A warehouse/mart is a means of *physically integrating data* from several sources, by bringing it all into one place. The advantage of physical integration is simplicity and run-time efficiency. Sometimes, however, physical integration may not be politically feasible because individual institutions in a multiinstitution consortium feel the need to retain "control" over their data and access to it, and believe that they will give away the store if they let people from other institutions access it directly.

In this case, virtual integration or "federation" is used. Software at a central site acts as a "mediator" between users and the data, and farms out end-user queries to one or more individual institution's data repositories—more accurately, "public" subsets of each. (Data such as financial information or leading-edge research data with intellectual-property implications, eg, would normally be off limits.)

Such an approach incurs needless performance penalties. This is especially true when different institutions use different DBMS technologies, different data models, and different means of encoding similar or identical data; cross-database querying then becomes needlessly complex (and limited). For such queries to work and return useful results, the individual sites should have put in a major effort to conform their individual data models to a "global" data model as much as possible. In practice, the collaboration between sites is rarely as intensive as it needs to be: the paranoia that prevents groups from moving "public" subsets of their data to a single physical shared site also inhibits collaboration.

The result is that the "global" data model tends to be a least-common-denominator model where the fine details of individual database schemas tend to be lost. The end result may turn out to be of minimal scientific value or relatively trivial. While an end-user may get the full details of an individual item that resides only on a single site, she/he may be out of luck for queries asking for a particular kind of data across multiple sites.

Federation should therefore only be used as a last resort, when physical integration is ruled out. National sponsors of research efforts who desire the generated data be a public resource are best advised to insist up front that awardees should transfer their data, once a particular "embargo period" has expired, to a shared physical resource. (The embargo period allows the discoverers to work up their initial findings and establish intellectual ownership before making their data public—this period is typically 6 months to a year.) This is what the National Human Genome Research Institute did with the awards related to the Human Genome Project, even occasionally threatening to pull funding from groups who were unreasonably tardy in submitting their data to Genbank.

> **Interestingly, I've seen one case in which a group tried to set up a federated database to multiple sources that all existed *within their own institution*, even though no political or technical barriers to physical integration existed. The only way to explain this was that this group didn't know, and hadn't bothered to learn, the first thing about data integration principles.**

9.3.2.3 Warehousing will not make custom reporting and data extraction go away

Data marts/warehouses are intended for end-user access and, in principle, to emancipate them by giving them free access to institutional data. Paradoxically, however, after warehousing is done, you may find that custom reporting and data extraction requirements actually *increase*. As Greenfield points out, this is because the more that end-users see what is possible with all the data in one place, the more they want. However, they don't always have the time or skills to do it themselves: they prefer to have queries set up for them that can be run regularly as reports. Also, while query tools may be reasonably intuitive, for queries of sufficient complexity, an expert who knows query languages and analytical tools must be called in.

In the case of EHR-derived data, specific considerations apply. No query tool, however intuitive, will automatically let users understand how the data is organized: acquiring this understanding requires time and intellectual effort. For example, users may expect to see fields called "hemoglobin" or "autism," not understanding that one is found as a data attribute under "laboratory tests" and the other one under "diagnoses." Also, queries may need to be rephrased. For example, the EHR considers an "encounter" as an individual interaction between the patient and a healthcare provider who could be a physician, a nurse, a blood-draw technician, etc. In the inpatient situation, however, users think of an encounter as an individual period of hospitalization, which translates in the EHR to hundreds of encounters.

9.3.2.4 Chronological and accuracy issues with EHR and hospital administrative data

All the different types of data do not go back equally in time. In institutions that have recently acquired a high-end EHR, selected legacy data from the previous (often homegrown) EHR may be imported into the current EHR, but the fine details of the legacy data are missing. For example, the details of individual encounters go back only as far as the go-live date of the new system. Procedures and laboratory data, however, may go back over several decades, but the details of the encounters involved are usually completely lost: the best one can do is utilize the associated date information.

Demographic data may go back even further, but unless the institution's hospital IT group has been employing the Social Security Administration's Death Master Index to update their records, or the US Post Office's database to update their address information—and currently, only a minority of IT departments do this—you may discover 120-year-old patients who, as far as the EHR is concerned, are still alive. My group recently used the EHR to identify and contact a large number of patients who had been seen in the *past year*, and we found that about 5% of their addresses, as recorded in the EHR, were out of date.

Another source of inaccuracy is related to the vagaries of coding.

- The fact that something is coded does not necessarily mean that the finding corresponding to the code actually occurred. Using administrative/billing data as a surrogate for clinical data may be problematic. In a study of data derived from the Veterans Administration database, Stein et al. [8] compared coded diagnoses of pulmonary embolism and abdominal hernia with the text notes. It was found that the code often indicated suspicion rather than confirmation of the condition; sometimes there were outright errors, where a subsequent negation of the diagnosis was miscoded as affirmation. In a few cases, there were negatives (ie, failure to code) where the diagnosis was apparent in the "postoperative complications" text, but not encoded.
- The same treatment can be coded in different ways. Certain hospitals, notably Columbia/HCA, have been cited for the practice of "upcoding"—selecting the code that will yield the highest reimbursement, even fraudulently upgrading the seriousness of a patient's medical condition to achieve this [9]. Until the feds cracked down, this practice appeared to be widely prevalent across healthcare organizations: HCA had to pay a $95 million fine [10]. (Upcoding is somewhat of an art, and if done cautiously, can even be legal when the patient has multiple conditions, just like maximizing one's tax deductions by listing them in one place rather than another.)

Detailed documentation must therefore accompany the warehouse or individual marts. Along with descriptions of the data elements related to each category of data, one must also document how far back it goes and the limitations in quality. (Limitations are especially important because end-users may naïvely treat the contents of a warehouse as gospel, whereas the data is only as accurate as the source: data such as nursing flowsheets may be particularly problematic, containing abundant narrative text in fields that were supposed to be numeric.) Such descriptions also constitute deliverables—their quality must be beta-tested and maintained to ensure that descriptions are current.

9.3.2.5 Redundancy and pseudoredundancy within the EHR

Sometimes, information will be recorded in two different places in the EHR: BMI, for example, may be recorded in vitals and also (in inpatient settings) in nursing flowsheets. Sometimes two different but closely related parameters will be recorded under the same name erroneously. Thus, birth weight is recorded when a child is born, and the child is weighed again when transferred to neonatal intensive care. This latter weight is some-times wrongly labeled "birth weight," which it is not: between the time of birth and the time of arrival in neonatal ICU, an intervention (eg, parenteral fluid administration) may have been performed, or an event (such as fluid loss) may have transpired that causes the two weights to be measurably different.

Your ETL process may have to rely on deep knowledge about the way your particu-lar EHR works to deal with redundancy when you organize your data, or distinguish between two parameters wrongly named the same.

9.4 CLINICAL REGISTRIES

A registry is a secondary database that records data on a specific group of patients. Regis-tries are a special case in that a considerable proportion of the data may be entered man-ually through abstraction of the EHR by domain experts. Manual abstraction, while ob-viously not the preferred means of information extraction, is often unavoidable because the reporting requirements served by registries are highly specific, and the information to be recorded must be synthesized by human expertise through review of several parts of the medical record. Sometimes, synthesis is necessary because the system from which the original data was captured recorded the parameters only as narrative text.

> For example, in institutional cancer registries that capture data to be submitted to NAACCR (the North American Association of Clinical Cancer Registries), there are fields that records how the diagnosis, or grade of the disease, or extent of spread, was established: one of the choices is "Determined at au-topsy." While there is a standard template for autopsy information, much of the fine detail continues to be captured as narrative text because of the sheer diversity of findings that may be encountered, related to a host of causes of natural or unnatural death. For example, toxicology examination is performed when poisoning or drug overdose is suspected, but not routinely in a termi-nally ill 90-year-old who died over a couple of months in hospice care.

Some registries, such as those for notifiable infectious diseases, are maintained by state governments and feed their content into larger national registries. Others must be maintained by individual institution as part of the contractual agreement with national sponsors: thus, any designated Comprehensive Cancer Center must maintain a local can-cer registry. Yet other registries are initiative specific—such as the various local/national registries used for specific organ or tissue transplants. Finally, many institutions with specific research interests may maintain local registries for conditions for which national initiatives may not exist.

Depending on the size of the initiative, a registry may be *distributed* (ie, each center maintains data on its own patients) or *centralized* (every group sends data to a central location). Usually, centralization coexists with local copies. Where multiple national sites are focused on the same condition, with each site collecting data on patients in their vicinity, a standard data format is usually employed. This means that everybody collects the same data elements and records them in an identical fashion, using the same encoding.

Certain EHR vendors misapply the term "registry" for canned queries and reports (typically based on diagnosis codes) that retrieve data in the EHR for patients with particular conditions, such as diabetes or heart attacks. Such data may be a useful starting point for creating a registry of your own, but don't fool yourself thinking that this raw data will meet reporting requirements for national registries focusing on the same condition. One of the problems with the built-in "registry" functions of EHRs is that vendors rarely make it easy to customize the queries or reports so that you can modify the criteria involved in identifying patients, or limit (or expand) the data on the patients that you retrieve.

9.4.1 Archaic data formats

Some registries go back to the Hollerith punched-card era, before the use of electronic computers, and their formats (unfortunately) still reflect this. For example, the NAACCR format for cancer registries uses fixed-width text where, say columns 48–50 represent a particular field, and columns 250–251 something else. Obviously, end-users who have to enter abstracted data would be driven crazy if they worked with such a format directly. Therefore numerous "registry" software packages exist that use a conventional microcomputer relational database for data entry purposes and export data into NAACCR format on demand.

However, for analysts who have to work with historical data, the work is made unnecessarily complex by the continued existence of such formats. For example, since a patient can have multiple races, NAACCR records these as five fields, Race1, Race2, Race3, and so on: some fields go all the way to number 26. In other words, the data is not in first-normal relational form, where repeating groups are *verboten*, being moved to separate tables where the numeric 1, 2, 3, etc., suffixes are eliminated. While splitting a giant line of 5000+ characters of text into individual fields is not rocket science (free utilities to do this exist), the "flat," nonrelational structure has to be restructured into a more logical, relational structure (into multiple tables) to allow query using modern tools.

Another consequence of the archaic format is that sometimes, the same fields end up doing multiple duties. Thus, one set of fields in the NAACCR format relates to "site-specific factors," where the semantics of a particular field depend on the site/anatomical location of the patient's primary cancer. That is, if the record describes a breast cancer, the field means one thing; for prostate cancer, it means another; for ovarian cancer, yet another. Any database-design student today who designated multiple-duty fields would either be flunked or made to perform 50 finger-pushups: multiple-duty fields immensely complicate data query for site-specific factors (such as identifying all patients who are estrogen-receptor positive for breast cancers) without consulting the NAACCR documentation and then writing case-by-case code. During the dark ages of data management reflected by this format, however, multiple-duty fields were about the only way to prevent unpredictable growth of the file structure, which circumstance would presumably give punched-card operators heart attacks.

> **Interestingly, I've seen cases in which the raw NAACCR format was ported directly to a relational data mart *without change*. This made the data impossible to query by end-users, who were constantly forced to refer to the 300 page+ NAACCR documentation to make sense of the data. This exemplifies an extension of Edsger Dijkstra's dictum that bad code can be written in any programming language: unusable databases can be similarly designed with any DBMS technology.**

9.5 ENCODING DATA PRIOR TO WAREHOUSING: STANDARDIZATION CHALLENGES

In addition to supporting the query/reporting needs of your institution's own end-users, warehouses/marts serve another purpose: interoperating with other institutions' corresponding systems, typically at the level of aggregate data sharing or sharing of (de-identified) individual patient-level data. Such sharing may be performed as part of Health Information Exchange (HIE) or for collaborative research. For data to be sharable, it is necessary to map data elements to standard vocabularies.

For diagnoses, this is not an issue, since everyone uses ICD-9 (or ICD-10) codes. For other clinical parameters, however, the use of standard codes may simply not be enforced in the EHR itself. Incorporating such codes can constitute a major curation effort.

- In most EHRs, entities such as lab tests, medications, or procedures have a local ID—an auto-incrementing integer (which is internal to the institution's EHR and meaningless beyond it) and an associated name. The set of names forms a "local" vocabulary. However, such vocabularies are hardly "controlled": very often, you will find that other descriptive/documentation fields are minimally populated.

  Further, to fully describe a lab test result, there are fields such as data type, tissue/fluid employed, units, maximum and minimum permissible values, maximum and minimum range of normal, set of permissible values (if categorical), and so forth. Consider yourself lucky if more than 10% of your lab-test descriptions are fully documented. You cannot necessarily rely on automated matching of names for mapping: for example, "potassium" in English becomes "Kalium" in German, and some institutions may use the chemical symbol "K." Some test names may not be fully descriptive: I've seen the test for serum triglycerides documented only as "TRIGS," with no other description anywhere else in the record. Such an abbreviation is hard to interpret in isolation: few users searching for "triglycerides" would be able to locate it easily.

  Also concerning, certain EHRs will not enforce the most basic integrity checks, such as ensuring that all lab-test names are unique, or forcing certain documentation fields to be nonblank. (In other words, duplicates and other errors may creep in, and you don't discover how messy the definitions are until someone decides that a curation effort is necessary.)

Mapping laboratory tests to LOINC becomes laborious because the documentation necessary to do the mapping accurately is often missing from the EHR, as stated earlier; you have to gather the information through interviews with the laboratory staff. LOINC also has its idiosyncrasies: if a test for a substance is expressed in different units (eg, pounds vs. kilograms; milligrams vs. micrograms) its LOINC code does not change *except when* it is expressed in molar units (eg, micromoles per unit volume). This does not make sense from the scientific perspective: a mole is merely a different kind of unit, based upon the substance's molecular weight in grams. In other words, mapping to LOINC is necessary but not sufficient for interoperation: all systems that share a particular test must have the data converted to an identical unit at some point. Otherwise, if pooled, "normal" numerical values may vary by several orders of magnitude.

- Procedures tend to be recorded in an even more chaotic fashion. While a standard vocabulary exists—the Healthcare Common Procedure Coding System (HCPCS), which is an extension of the American Medical Association's Current Procedural Terminology (CPT)—this is not the easiest for healthcare providers to use, and so a local vocabulary is often employed. Here, there is often no clear rationale for adding new procedures to the system. I've seen entries that combined the type of surgery (eg, cataract) with the name of a particular surgeon. (And if that surgeon leaves the institution to go elsewhere, that entry becomes clutter.)

  A small army of "coders" in the billing department have the job of reading the clinical notes along with the procedure-code information, talking to individual doctors if clarifications are needed, and then manually creating HCPCS codes in the billing system. I've also discussed the problems with secondary coding in Section 9.3.2.4: however, if you wish to make your procedure data usable, you have little choice other than to use the HCPCS-encoded billing data.

- The way individual EHRs manage medication data varies widely. For example, in many EHRs, daily dosage is not a discrete numeric field but recorded as narrative text (1 tab 6-hourly). You have to compute the dose using some form of NLP that parses the text instructions and combines them with the tablet strength, which is also buried in the narrative text of the medication name. Mapping to RxNorm gets very complicated if you try to map at the level of preparation: you may save much labor if you map to the generic drug only.

  In general, you need to use one of the medication knowledge bases that records hierarchical information linking medication classes (eg, angiotensin inhibitors, cardioselective beta blockers) to individual generic medications, which are in turn linked to the individual medications/brands tracked in the EHR. To complicate matters, the relationship between generics and individual medications is many-to-many because certain medications are combinations of one or more generics. End-users rarely care about individual medications or preparations: their queries are either at the generic drug level or the drug family level.

In summary, local EHR vocabularies tend to be an example of the Second Law of Thermodynamics applied to databases: the level of disorder in the system tends to increase over time. Further, if individual clinical parameters are not adequately documented in the EHR, end-users will not be able to locate them readily when querying the data if you simply take raw, uncurated EHR data and warehouse it—as I've seen more than one group do. Mapping to controlled vocabularies has the benefit that the detailed documentation of an item in the controlled-vocabulary source is inherited through the mapping so that the usability of the mart or warehouse improves dramatically. This improvement can be so significant that, even if your data will *not* be shared with other institutions, you should still consider mapping your data elements to standard vocabularies.

Disorder is invariable in all user-defined vocabularies where curation efforts are lax or nonexistent. Certain features, such as LAYGO (Learn As You GO), a feature of VAMC's Vista EHR, predispose it to disorder. Here certain pull-down/list fields are based on predefined sets of items (which may be stored in other tables). In LAYGO, when a user types into the field, if the entry does not exist in the set, the system asks if he wants to add it to the set. If the user responds "Yes," the set grows by one item and is available immediately to all users.

This seems like a good idea at first sight. However, spelling mistakes and synonyms may also be accepted as new entries: a spelling-challenged, lackadaisical user who can't bother to check what was typed when prompted can be the equivalent of a Typhoid Mary to an institution's local vocabulary efforts. In the case of Rite-Aid's pharmacy software, LAYGO resulted in some twenty alternative spellings of acetaminophen, each of which was treated as a new drug and assigned a new ID. This set back their warehousing initiative until all the garbage was cleaned up by an expensive manual curation process. I strongly recommend that if you can turn off your EHR's LAYGO feature, do so: you'll save a fortune in headache medication for your team members.

9.6 RELATIONSHIPS BETWEEN HEALTHCARE IT AND HEALTH INFORMATICS GROUPS

I pointed out earlier that in many or most teaching institutions, the clinical research informatics group and hospital IT do not necessarily report to the same ultimate authority. The informatics group is usually employed by the university (and led by faculty), while IT is employed by the hospital. This is not an ideal situation and can impact research as well as quality efforts. Ideally, the two groups would operate as one, with shared goals, and would see eye to eye on priorities. In reality, I've seen relationships varying from the truly harmonious, to an uneasy truce, to outright hostility where one group wouldn't give the other the time of day.

Dysfunctional relationships can seriously impact research.

- I know of one case where the hospital's chief information officer took 2 *years* to grant access to the EHR's relational data store to the research team, so that creation of a research data mart was postponed inordinately.
- I've also seen a senior researcher who was outright abusive and developed a reputation for simply dumping work on the IT staff, making last-minute demands, constantly changing his requirements, and lacking the basic courtesy of a "please" or "thank you." After a while, the IT staff simply decided to become passive-aggressive and slow-walk all requests. Unfortunately, other researchers also suffered through guilt by association: they now had to make extra effort to be nice to the IT staff to undo the damage their colleague had done.

- Initiatives intended to utilize the EHR's personal-health-record facility to contact patients (eg, for recruitment or as part of a longitudinal study) may not be turned around as quickly as desired, because the staff responsible for setting up EHR-based custom messaging for a study does not report to the research team and in fact may have a bunch of other responsibilities. Similarly, in warehousing efforts, problems discovered by the research team with local vocabularies (like problems with all data) should ideally be fixed in the source. Without a close collaboration, however, the research team will need to apply the band-aid of fixing problems downstream, so that the root causes, which make the errors keep (re)appearing, may never get fixed.

There is no magic solution here: pissing contests between rival bureaucracies have probably existed since the time of the ancient Egyptians, and problems that have been festering for years will not go away overnight. The only advice that I can give the reader of this book is not to be discouraged by encountering an apparent wall of resistance, and not to confuse an organization or bureaucracy with the individuals within it. While some people are pathological, unredeemable hard cases, others are just average folks trying to do their jobs, and treating them with respect—and if necessary, going out of the way to help them or do them little favors that you wouldn't be expected to—can go a long way in letting you get things done under the radar. Even if they resent your bosses, they may regard you as one of the "good guys" and reciprocate. The challenge, of course, is in identifying who to cultivate and who to abandon as lost causes.

BIBLIOGRAPHY

[1] T. Halpin, Information Modeling and Relational Databases: From Conceptual Analysis to Logical Design, Morgan Kaufman Publishers, San Francisco, CA, (2001).

[2] P. Nadkarni, Database normalization (BIS560 lecture notes). Available from: http://ycmi.med.yale.edu/nadkarni/db_course/normalization_frame.htm, 2015.

[3] S.N. Murphy, G. Weber, M. Mendis, V. Gainer, H.C. Chueh, S. Churchill, I. Kohane, Serving the enterprise and beyond with informatics for integrating biology and the bedside (i2b2), J. Am. Med. Inform. Assoc. 17 (2) (2010) 124–130.

[4] W. Inmon, K. Rudin, C. Buss, R. Sousa, Data Warehouse Performance, John Wiley & Sons, New York, (1998).

[5] R. Kimball, L. Reeves, M. Ross, W. Thornthwaite, The Data Warehouse Lifecycle Toolkit, 2nd ed., Expert Methods for Designing, Developing, and Deploying Data Warehouses, John Wiley, New York, NY, (2008).

[6] R. Kimball, J. Caserta, The Data Warehouse ETL Toolkit, Wiley Computer Publishing, New York, (2008).

[7] M. Demarest, The politics of data warehousing. February 2, 2016. Available from: http://www.uncg.edu/ism/ism611/politics.pdf, June 1997.

[8] H. Stein, P. Nadkarni, J. Erdos, P. Miller, Exploring the degree of concordance of coded and textual data in answering clinical queries from a clinical data repository, J. Am. Med. Inform. Assoc. 7 (1) (2000) 42–54.

[9] Lagnado L. Investigators probe upcoding and hospitals' profits from it. The Wall Street Journal April 17, 1997. Available from: http://www.wsj.com/articles/SB861227444157204500, 1997.

[10] Eichenwald K. HCA to pay $95 million in fraud case. The New York Times December 15, 2000. Available from: http://www.nytimes.com/2000/12/15/business/hca-to-pay-95-million-in-fraud-case.html, 2000.

CHAPTER 10

Core Technologies: Data Mining and "Big Data"

10.1 INTRODUCTION

Data mining and "big data"—on which mining techniques are employed—are "hot" topics, in part because of the hype surrounding them and the extravagant promises that are being made on their behalf. And this is not just in the biomedical realm: the thirst of the intelligence agencies for massive budget increases to gather ever more surveillance data is well known, though whether commensurate improvement in readiness against terror attacks has resulted is questionable. In this chapter, I point out both the promises and limitations of the technology. I'll start by defining each term rigorously.

Rigor is necessary to counter the tendency of the ethically challenged (who are unfortunately also the most publicity-seeking) to redefine these terms to fit whatever product, approach, or activity they happen to be selling, advocating, or performing. As an egregious example of snake oil, it is hard to top the Information Awareness Office (IAO) [1], which operated from 2002–2003 until shut down by Congress. For details, read the previous reference, as well as the Wikipedia article on "Policy analysis market" [2], a proposal that ended the second career of IAO's director, John Poindexter, who had been previously convicted for his role in the Iran-Contra scandal. Even the data-mining textbook of Leskovec, Rajaraman, and Ullman [3], to which I refer you later, points out, in the very first chapter, some of the fallacies of the IAO's "Total Information Awareness" program, and uses it to illustrate the Bonferroni correction for multiple hypotheses, introduced in chapter: An Introduction to Clinical Research Concepts.

Data Mining is an exploratory data-analytic process that detects interesting, novel patterns within one or more data sets (that are usually large). It employs a variety of techniques, including the machine-learning techniques of chapter: Core Technologies: Machine Learning and Natural Language Processing and standard multivariate statistical techniques.

Big data is data that has been defined by Snow [4] as having all of the characteristics defined by the "4 Vs": volume (lots of data), variety (highly diverse data), velocity (changing very fast), and veracity (hard to fully validate). I'll elaborate on these aspects later.

The science underlying the field is vast. In the modest space allocated here, I can do little more than emphasize a few fundamental principles, which will hopefully let

Clinical Research Computing. http://dx.doi.org/10.1016/B978-0-12-803130-8.00010-5

you appreciate the sources listed at the end of this chapter. (The sources list also serves as additional reading for chapter: Core Technologies: Machine Learning and Natural Language Processing, because of the close tie-in to machine learning). Easily the most accessible book is Nate Silver's classic, "*The Signal and the Noise: Why so many predictions fail – but some don't*" [5].

If you haven't read Silver's book, you should probably put my book down, reserve a copy from your local library or order it for your e-reader, and then come back. Silver is a statistician/political analyst who became famous for predicting the results of the 2012 US presidential election accurately for all 50 states, using a meta-analytic method that pooled several independent poll results to eliminate bias [6]. His book provides a panoramic view of "big data": not only the few successes, but the far more numerous failures.

10.1.1 The four Vs of big data

I now elaborate on each of the four Vs of big data.
- *Volume*: Today, "large" means on the order of terabytes or more.
- *Variety*: This implies a great diversity of data types: a mixture of structured, unstructured, and semistructured data, coming from a large variety of sources, and consequently in a large number of formats. Variety is related to the issue of "polyglot persistence," introduced in chapter: Core Informatics Technologies: Data Storage, Section 3.5: no single DBMS or data processing technology will suffice to manage all of the data.

    Some systems may employ RDBMSs and be oriented toward serialized transactions that emphasize consistency above all else. Other systems may employ "NoSQL" technology that maximizes throughput, even at the cost of temporary inconsistency. Yet other systems may be nonvolatile data sources that are not interactively updated at all—such as data pulled using web-crawler software, which may be updated once a day, or data in analytical databases whose contents are derived from other (primary) systems. In any case, trying to convert all data into a single format is a quixotic and futile effort: it is best to employ multiple formats—even for the same logical data—each best suited to its intended purpose.
- *Velocity*: Rapid growth is both in terms of volume, as well as in variety. In other words, new types of data may be continually introduced while new data sources are incorporated. This is especially true of data that is being gleaned from sources such as the web.
- *Veracity*, or rather, the lack of it: When data comes from varied sources, the quality of data in each source can vary greatly—some data is very clean and well-curated, other data sources are one or two levels above "garbage." Information on the same fact, as recorded in two or more different sources, may conflict with each other.

Data of very poor quality—which often occurs when you have no controls put in place during the data-gathering process—can be more trouble than it is worth, and its incorporation into a larger set can have an effect best described by the earthy metaphor "pissing in the soup": the dataset as a whole now becomes unconsumable. Often, skill in "big data" science is identified by good judgment in trying something (including a new data source) in pilot mode, carefully quantifying investment versus benefit, and cutting your losses and walking away in case of failure.

Borderline-quality data may be salvaged through data-cleansing methods, but these methods must be algorithmic, that is, employing computer programs where you hope that the false-positive and false-negative error rates are modest. There are insufficient resources available to curate the data manually: the data is coming in too fast for that to happen, and there is far too much of it. Consequently, you have to be happy with data that is 95% accurate, or even 90% accurate.

The effort required to move from 95% accuracy to 99% accuracy is dispro-portionate to the gain in quality—and to move to 99.9% accuracy is more challenging still. While the semiconductor industry achieves incredible purity levels for silicon (less than 1 part of impurity per 100 billion), data cleansing is much more challenging because data created by fallible humans is less con-trollable and messier than computer-controlled industrial processes.

Quality is a function of purpose. For Google's search algorithm, for example, it does not matter that the viewpoint of a page that is being served up is pure intellectual swill: all that matters is that the page be relevant to a user's search—even if many users of Google happen to be ignorant or bigoted. Try Googling "Flat Earth Society" or "Creationism," and you will get pages dedicated to advocating these viewpoints very high up in your search results. If your business relates to geographical navigation or to biological science, however, using these pages as "information resources" would set your efforts back considerably. If you were creating a bibliography of various false beliefs, however, they might be useful.

Managing "big data," and being able to *report* on that data with the hope of generating actionable insight, is so challenging that you find yourself in a variant of the "three-legged-stool" problem. You can get your reports in a timely fashion, you can implement your reports with a moderate investment in human and/or hardware resources, or the numbers in your reports can be highly accurate—pick any two out of the three.

However, it is fair to say that no application in the field of clinical research informatics, as of today, meets these criteria. We are way behind Internet-based industry or the physical sciences in terms of both the size of the problems that we handle and our current skillset. This may change, however, if sensor data from mobile devices, along with large-scale genomic data, start getting incorporated into the medical record.

In this regard, many EHR systems use pre-RDBMS technologies that are simply not designed to store large volumes of unstructured primary data. Even data such as electrocardiographic traces, or static or dynamic radiology/pathology images, are stored in separate systems, and only summary information, textual or semistructured, is transmitted to the EHR.

Additionally, few EHR implementers seem to know the first thing about rep-resenting genomic data, even at the basic single-nucleotide-polymorphism level: to most EHRs, a given polymorphism is simply a lab test, like a hemo-globin level, with the result recorded as text. If personalized medicine ever

becomes routine a decade or two from now, such EHRs are likely to collapse under the sheer weight of their unwieldiness, unless they undergo a drastic redesign.

10.2 AN OVERVIEW OF DATA-MINING METHODOLOGY

Large-scale data gathering and mining of the resulting datasets led to the surmise by Anand Rajaraman [7] that "more data usually beats better algorithms." While this claim has been vigorously disputed [8,9], the fact is that many algorithms (such as those based on N-grams, discussed in chapter: Core Technologies: Machine Learning and Natural Language Processing) require large sets in order to perform reliably.

> There is certainly no barrier to the development of new algorithms that incorporate deeper computational insights *and* utilize copious data for learning: this is how hidden Markov models and conditional random fields made their mark. Further, Rajaraman himself seems to have made this statement just to provoke. As co-author of the excellent data-mining text cited earlier [3], he and his co-authors repeatedly illustrate how a simple approach to solving a particular problem gets progressively improved algorithmically by various researchers, or even the same group, to be more efficient or accurate, or how it gets modified to deal with varying circumstances. For example, Google continually tweaks its algorithms to thwart those who try to fool its PageRank approach.

10.2.1 Big data and Bayesian statistics

I introduced you to Bayes' theorem in chapter: Core Technologies: Machine Learning and Natural Language Processing. It is important to remember that the probabilities that are computed *are not fixed and may change over time*: thus, prevalence of a disease may change, and the clinical presentation of the disease may change with respect to symptoms and severity. For example, the polymath Jared Diamond, in his classic, *Guns, Germs, and Steel* [10], notes that when syphilis was first brought into Europe from the Americas via Columbus's sailors (it was prevalent in the Native American populations of the Caribbean, who were affected relatively mildly and invisibly), it was devastatingly lethal to its European victims, killing them in a few months.

Big data—specifically the availability of large datasets accumulated over long periods—makes Bayesian methods much more robust because it enables estimations of probabilities to be much more reliable because of the enormous sample sizes involved.

> Silver, however, points out that probabilities may need to be revised within a short time rather than over years, using as an example the crashing of two airliners into the World Trade Center towers on 9/11. While the first crash had an extremely low probability according to Bayes' formula, once it *had* occurred, the probability of the *second* crash was extremely high. Silver also cites an example of

a highly successful professional (Bayesian) gambler who continually updates his calculations about what (or what not) to bet on, based upon new information.

A phenomenon pointed out by Provost and Fawcett [11] is that sometimes the Bayesian probabilities change over time because of software implementations that use the probabilities—a sort of self-canceling phenomenon. An example is e-mail spam: in response to the anti-spam measures in e-mail software (which utilize Naïve Bayes, as mentioned in chapter: Core Technologies: Machine Learning and Natural Language Processing), spammers have also been getting more clever and avoiding the obvious terms that trigger the spam filter.

10.2.2 Data mining problems and techniques: principles

Many of the data-mining techniques (map-reduce, regression techniques, support-vector machines, naïve Bayes, clustering, information retrieval, the use of NoSQL technologies, etc.), have already been covered in the Core Technologies chapters: Core Informatics Technologies: Data Storage and Core Technologies: Machine Learning and Natural Language Processing. New techniques are being developed continually. While the traditional problems of machine learning are classification and regression (and clustering and dimension reduction for unsupervised methods), Provost and Fawcett point out that data mining techniques additionally focus on one or more of the following.

- *Similarity*: Finding out what items are "similar" to others. This forms the basis of recommendation systems; when applied to biological sequences, it is the basis for discovery of either similar function or similar origin.
- *Co-occurrence*: Which items co-occur: the co-occurrence could be items purchased/viewed together (or by the same person), or words that occur in the same abstract/sentence.
- *Profiling*: Characterizing the typical behavior of an individual in a population so that we can flag anomalous behavior. (This is the principle behind fraud detection.)

Some of the problems that are addressed are essentially the same as those addressed by traditional multivariate statistics, except that the size of the problem has increased a hundredfold or more. In many cases, trying to obtain exact answers would be prohibitive in terms of either computer resources or time. Also, the algorithms used previously assumed that the data to be processed fit completely into memory (eg, traditional clustering algorithms), which may not apply anymore.

Therefore, to address large-scale problems, algorithms need to be overhauled. There are several types of approaches employed.

- One approach employs *probabilistic sampling* methods, using random samples of the data rather than all of it to perform analyses, and then applying standard statistical theory to the numbers obtained to estimate the confidence limits when extrapolated to the entire data. This approach is typically used for streams of data (especially from natural phenomena) when the data arrives so fast, and there is so much of it, that there is no time to transfer all of it to memory, and we need real-time answers.
- Another approach, used to shorten computational time, is to develop *fast-approximation* methods that incorporate shortcuts and give approximate answers that have an acceptable

rate of error, but work very quickly, as opposed to exact methods that would take forever. Fast-approximation methods have been an active area of research in biological sequence comparison, given the enormous size of sequence databases.

- Finally, one uses *parallelism* on distributed hardware (ie, map-reduce tasks running on Hadoop clusters) to speed up computation. Parallelism is not employed for real-time processing, because it relies on dividing up the input across machines. Assuming, however, that the data can be stored on disk and real-time reporting is not a concern, this approach can be used. It may still be combined with approximate methods, because while hardware only allows you to scale linearly, many problems scale supralinearly in complexity.

10.2.3 Selected data-mining problems

Rather than try to cover every data-mining problem in depth (I refer you to books that do that very well, at the end of this chapter), I'll try to give you a flavor of the problems that must be solved. Examples are provided below. Specific topics that have not been covered in chapter: Core Technologies: Machine Learning and Natural Language Processing will be covered here.

- *Clustering data that is too large to fit into RAM,* such as social networks that consist of millions of people identified by unique IP addresses that access a large website.
- *Dimension reduction:* With dimension reduction, introduced briefly in chapter: Core Technologies: Machine Learning and Natural Language Processing, we hope to derive a simpler model (compared to the raw data) that explains *most* of the phenomena we are studying. A model becomes simpler either when we discard variables that have little predictive value, or when we derive (fewer) composites of the original variables that explain most of the variation in the data. I've already discussed principal components analysis, which was first applied in the field of intelligence characterization. I'll discuss another method, latent semantic indexing, in the next subsection.
- *"Association rules"* and *"market-basket" analysis:* An association rule is a prediction of the form, "If X, then Y." For example, the association of specific single-nucleotide polymorphisms with specific disease at a frequency much higher than expected by chance may suggest an etiologic relationship that makes the disease more likely.

  Market-basket analysis, an application of association rules, is used to determine what other items are commonly purchased by people who buy a particular item. For example, beer purchasers tend to buy savory snacks, so we could improve the shopping experience by placing them closer together in a store. This also allows cross-promotions.

  In biomedical NLP, one can look for "co-occurring concepts" in the same document (more usefully, the same paragraph or sentence), which may imply a higher-level joint concept or some kind of relationship between them. The National Library of Medicine has done this, producing a list of all pairs of concepts that co-occur in Medline abstracts.
- *Recommendation systems:* These are used by online vendors who have millions of items in virtual inventory (as opposed to "brick-and-mortar" operations who only sell thousands). These vendors make a living through what is called the "long tail." That is, a histogram of the number of each item sold, from most popular to least popular, extends very far to the right, that is, it is a long-tailed distribution.

  People go to an online vendor like Amazon or Netflix because they can locate an item published decades ago or an old movie. Their virtual inventory is far too numerous to list in a catalog, and therefore, if a customer buys/watches a rare item, there may be other items that are highly similar to it in content, also rare, but which that customer may not even know about. A recommendation system tells you that people who purchased or viewed item X also purchased Y, Z, etc., or that the customer may like other items (based on one or more measures of similarity such as genre, author, actors, etc.).

The real advantage of a recommendation system is that it can be customized to a single customer algorithmically at very low incremental cost-per-customer. Also, the more data repeat customers provide the system, the better the recommendation system gets, because it develops a "profile" of that customer. (Note that a customer doesn't have to fit a single profile—she may enjoy multiple movie genres, for instance.) For a physical retailer, it would not be cost effective to advertise a book that had only, say, 50 possible customers.

10.2.3.1 Latent semantic indexing

A dimension-reducing approach that extracts "topics" or "themes" from text passages is called *latent semantic indexing* (LSI) (the term "indexing" is sometimes replaced with "analysis"). This information-retrieval technique is based upon the matrix method called singular value decomposition (SVD), which is one of the methods for performing principal components analysis as well. I refer you back to Section 3.3.6 of chapter: Core Informatics Technologies: Data Storage to refresh your memory if you need to.

A highly accessible description of SVD, intended for someone with knowledge of little more than high school algebra, is David Austin's article, available at the American Mathematical Society's website [12]. Though the article doesn't mention LSI, it does mention Netflix's competition for a program that could improve its recommendation system, which is related to LSI. (The Wikipedia article, by contrast, is heavily technical, and should not be the first thing you refer to.)

We start by taking a collection of N documents and extract terms from each document after discarding "noise" or "stop" words. Let us assume that there are M unique non noise words across all documents. (In practice, N can range in the millions and M in the hundreds of thousands.) We can now create a giant matrix of M terms against N documents with each cell containing a weight, which is the TF*IDF metric (see Section 3.3.6): a measure of how often the term occurs in the given document multiplied by a measure of the infrequency of the term in the entire collection of documents so that rarer terms get more weight.

We now perform the SVD procedure (using a computer program, of course) on this giant matrix and extract a series of numbers, called the *singular values*. With each singular value, one also gets a set of M numbers, the corresponding *singular vectors*. The algorithms typically employed for large-scale SVD usually extract the *largest* singular value first, the second largest next, and so on, so that we can stop when the values that we have extracted cumulatively account for a certain proportion of the variation in the data, for example, 90% or 95%. (There is a very simple calculation that lets us check how close we are.) Even with a million documents, in practice it is rare that more than 500–1000 values need to be extracted. In other words, we have reduced the original dimension of a couple of hundred thousand "terms" to about a thousand "topics."

Each singular value corresponds to a certain "topic" or "theme," while the elements in the vector correspond to a weighting of the individual terms in the document for that topic. Many weightings will be zero or vanishingly small, which means that this particular term does not relate to the given topic.

> **For example, suppose we process a collection of documents related to the environmental sciences and we find a singular-value vector where the non zero terms are "algal blooms," "nitrogen," "fertilizer," "phosphorus," "dinoflagellate," "oxygen depletion," "red tide," "saxitoxin," and "estuary." This would suggest a theme of "seawater pollution due to fertilizer runoff in rivers"—an all-too prevalent phenomenon that has decimated the fish populations (and the fish catch) in the states of the US that border the Atlantic. (The nitrogen and phosphorus runoff in fertilizer, on entering the sea, gives a boost to dinoflagellate algae, which then multiply so rapidly that they produce toxins that then kill fish by the thousands: the algae become so numerous that the pigment they contain gives the water a reddish tint.)**

LSI has numerous applications. It can be used to automatically classify documents into categories based on the topics/themes that they contain. As an extension, these categories can be used as an additional means of indexing the documents for more sensitive retrieval in response to user queries. This capability has been used in applications as diverse as online customer support knowledge bases, in grant-review processes to assist matching of grants with reviewers, and search for "prior art" in patent databases, that is, highly similar, previously done work that invalidates the patent under consideration, because the present patent fails to meet the pre-requisite of novelty.

LSI is especially useful in knowledge domains where controlled vocabularies have not been developed, that is, most of the sciences and humanities. While biomedicine is believed to have extensive controlled-vocabulary coverage, as evinced by the National Library of Medicine's Unified Language System Metathesaurus, this is only true for specific areas of clinical medicine, where controlled-vocabulary efforts have contributed to the UMLS. Fields such as the various imaging modalities used in radiology, leading-edge neuroscience, or most aspects of laboratory bioscience (other than genomics) are underrepresented in the UMLS, and LSI can be used to find higher-level relationships between groups of terms, as an aid to streamlining manual curation.

Another advantage of LSI is that the technique is independent of language (as long as an individual analysis is done on a set of documents using a single language), and can be used to find similar documents across languages. For this, you need to have a set of translated documents that have been precategorized (in one of the languages you are interested in). If you then run LSI against the database where all the documents are in the other language, and then check for categories where the translated documents appear, you can then discover documents in other languages that share the same categories. In other words, you have performed a cross-language information retrieval.

LSI has been used to detect polysemy, where the same term is used in multiple related contexts. Many terms in medicine, for example, refer to substances that play a physiological role and are also measured in blood or urine (where the substance name becomes a short-form for the laboratory test for the same substance). In terms of SVD, the same term turns out to have a significant weight in two distinct topics/vectors. (The drawback here is that LSI gets confused when there is polysemy within the same document, unless it is used within a single document in a predominant sense.)

10.2.3.2 K-Nearest Neighbors

K-Nearest Neighbors (KNN) is a standard machine-learning method that has been extended to large-scale data mining efforts. The idea is that one uses a large amount of training data, where each data point is characterized by a set of variables. Conceptually, each point is plotted in a high-dimensional space, where each axis in the space corresponds to an individual variable. When we have a new (test) data point, we want to find out the K nearest neighbors that are closest (ie, most "similar" to it). The number K is typically chosen as the square root of N, the total number of points in the training data set. (Thus, if N is 400, K = 20).

KNN is conceptually simple and has the advantage of being nonparametric. That is, the method can be used even when the variables are categorical—though if you are using numeric variables in the mix, it is best to standardize them (see Section 4.4 of chapter: Core Technologies: Machine Learning and Natural Language Processing) to eliminate differences in scale. The challenge is that when the number of data points is very large (eg, an online bookseller has millions of books), special methods must be employed to rapidly search the space and find the "most similar" items.

Usually, some form of precomputation is employed for example, indexing. In addition, rather than using all the data points, selected data points that are representative of individual clusters ("prototypes") may be used to facilitate the search against a new item, and then the precomputed neighbors of the most similar prototype are also displayed. Similarly, attempting to reduce the number of dimensions with a method like SVD/LSI and then plotting the data points in the reduced variable space may result in significant gains in performance.

The Wikipedia article on this topic [13] is well-written and highly approachable.

10.3 LIMITATIONS AND CAVEATS

Now for the "downer" part of the chapter. While some big-data applications have definitely made their mark in the business world, in other areas, the success has been less stellar. The record of the National Security Agency, which has put more resources into "big data" than any other organization, is more of a mixed bag. Some "old timers" have complained that with all the money being poured into technology, resources have been diverted from supporting the humans on the ground [14]. (NSA's algorithms and super-computers, for example, failed to foresee the rise of ISIS.)

> **Maybe healthcare will be an easier problem than security because disease pro-cesses, unlike terrorists, don't have minds of their own, but on the other hand, microbes are incredibly adaptable. Some fungi have even adapted to thrive in the abandoned nuclear reactor in Chernobyl, Ukraine, site of the infamous 1986 meltdown, using the residual, and still lethally high, gamma-radiation as a source of energy, just like chlorophyll-containing plants use sunlight [15]. So even if worldwide nuclear holocaust comes to pass, some form of life will be around.**

10.3.1 Hypothesis discovery versus hypothesis confirmation

It is very rare that a big-data/data-mining effort, *by itself*, will result in new discoveries. Most of the time, all that mining suggests is an *association*, but discerning cause versus effect, or even confirming the association, is rarely possible: luck is hard to systematize. This happens for at least two reasons.

1. Much data mining is *opportunistic*: That is, existing data previously gathered for an entirely different purpose is reanalyzed. In such a case, the critical parameters that are essential for confirmation of a hypothesis may not have been gathered, and therefore the volume of data is irrelevant. The reasons for missing parameters are either that nobody thought to gather them at the time, because no one had discovered the parameter then, or the parameters are too expensive or infeasible to gather at the time the observation was recorded.

   An example of finding the critical parameter happened in 1980, when there was an epi-demic of teenage girls and women coming in to emergency rooms with high fevers, red skin blotches, and in a state of shock. The critical breakthrough was when someone noticed that all the women had been menstruating. The problem turned out, on inquiry, to be associated with hyper-absorbent tampons, in particular Procter and Gamble's Rely brand, which facili-tated the growth of bacterial toxins [16]. The toxins then entered the body through vaginal ulcers (which the induced dryness predisposed to) and was responsible for the symptoms. Information about tampon brand is not normally captured in EHRs; a data-mining effort of retrospective data would not have struck gold.

2. Even when the parameter/s of interest may have been recorded, there is no guarantee that they have been captured while rigorously controlling the covariates that could influence the outcome; in other words, the findings may be coincidental.

Therefore, *prospective* designs are usually necessary. Even here, you can't be a pack rat and insist on capturing *everything*, without knowing what you will do with each item: you must have a clear hypothesis in mind. As Robert Califf points out in the context of large clinical trials [17], sometimes adding a single data point to a study can add hundreds of thousands of dollars to the budget, when that data point happens to be an expensive lab test. But even if the test involved did not involve high cost or patient risk, you still have to budget for the human resources to collect it—typically research staff.

For subjective findings, you have to ask the patient questions, and as I've mentioned in an earlier chapter, it is possible to wear out your welcome if your data collection gets too aggressive. But even if the patient were willing, don't expect overworked doctors and nurses to do this for you gratis: they are already drowning in documentation requirements. The assumption that data about everything that happens to the patient can be magically sucked up in vacuum-cleaner fashion and put into a usable form into an EHR or clinical data management system is not valid.

Similar concerns apply to "big-data" designs that utilize sensors or, in the case of efforts that utilize web data, automated web collection, where the costs of collection are dramatically lower on a per-datum basis. When big-data efforts succeed, it is because of careful thought and deliberate design as to what is captured. Thus, when Netflix records the time for which you watch a movie and then abandon it (never to watch it again), it is because this is a means of discerning your preferences and fine tuning their recommendation system.

10.3.2 Data mining and "knowledge discovery"

While data mining has often been hyped to the public as "knowledge discovery," some words of correction are in order. The discovery of patterns hardly ever constitutes "knowledge," a term that implies *actionable insight*. (As Silver points out, the perception that a wild animal is charging at you is *data*: it does not become knowledge unless you can do something about it.) Pattern discovery is but the first step, only yielding leads that must be followed up through other means.

I've already stated that prospective experimental designs are necessary for hypothesis confirmation. Very often, *multiple* experiments and observations must be performed, each directed toward solving a part of the puzzle, as illustrated in example 2 below.

The number of facts that constitute "knowledge" can vary from one to a very large number, and all the necessary facts are rarely available in a single data source.

1. Brilliant insights from observation of a single fact are very rare. One of importance in the history of espionage is described by Miles Copeland in *The Real Spy World* [18]. During World War II, prior to the 1944 Normandy invasion, the US conducted a series of bombing raids on French railroads to disrupt large-scale movements of German troops, who might be brought in from the south as reinforcements to the northern French coastline where the invasion was planned. A large network of allied spies, who were managed by the US Office of Strategic Services (OSS, the CIA's predecessor), had the task of determining whether individual raids were successful.

   However, according to Copeland, the network's reports almost always went into the wastebasket. An OSS analyst, Walter Levy, had found all that he needed to know by looking at the aerial photographs taken by US bombers after they finished their raids, and more importantly, by checking the price of oranges on the Paris market. The price went up after a successful raid disrupted their transport from Marseille to Paris, and plunged after repairs had restored transport. After the war, Levy became a world-famous analyst who was considered the "dean of United States oil economists" [19].

2. At the other extreme, the association of the global weather disruption of 1816 with the eruption of the volcano Tambora in Indonesia in April 1815, was proved relatively recently in the late 20th century. The year 1816, referred to in the US as "the year without a summer," was characterized by failed harvests as far away as Europe, India, and North America. Detecting the pattern itself required locating historical records from multiple locations.

The actual insight came from several sources: studies of the eruption site that estimated the volume of material released during the explosion, the discovery that atmospheric dust causes cooling, the discovery that sulfuric acid aerosols (formed by oxidation and hydration of the sulfur dioxide released from the eruption) cause a "reverse-greenhouse" effect through reflection of the sun's incident light, and through indirectly measuring spikes in atmospheric sulfate concentrations in 1816 through polar ice cores from sites as far away from the eruption site as Greenland and Antarctica.

10.3.2.1 The needle-in-the-haystack problem

When data mining has been applied to surveillance/security, the problem is that it is often not just a matter of looking for a needle in a haystack: it is that more often the mining exercise discovers numerous possible needles—and throwing out the false signals requires in-depth domain knowledge as well as a lot of time and human effort. The NSA missed 9/11—despite the reports the FBI received about the suspicious behavior of Zacarias Moussaoi, who was taking flying lessons but didn't seem to care to learn about taking off or landing the plane. In World War II, despite the fact that US intelligence had broken the Japanese codes, the Pearl Harbor attack still occurred.

10.3.3 Aside: data + data-mining technique = publication

Many data-mining neophytes in academia think of data-mining tools as a means of rapid publication: take some data, crank away at it with some off-the-shelf tools or a new approach devised by the authors, and describe the results obtained. Many old-time statisticians have therefore understandably confused data mining with the term "data dredging," also known as "torturing the data until it confesses"—a phrase attributed to the late 1991 Economics Nobel Laureate Ronald Coase. Dredging as a scientific activity comes in waves: the first wave began in the 1930s with the discovery of the Pearson correlation coefficient, as described in chapter: An Introduction to Clinical Research Concepts. In the second wave, following the general availability of computing power, stepwise multiple linear regression took its place, and now that large databases are commonplace, poorly-thought-out data-mining exercises represent the third wave.

I wish I had a buck for the number of manuscripts that have been submitted to research journals and scientific conferences all over the world that unfortunately omit the final, critical step of evaluating the reported findings in the context of the published literature to determine novelty and significance. Many of the submitted results yield insights that are not very far removed from the (possibly apocryphal) story of the data-mining program that, after ploughing through a cancer-patient database, reported triumphantly that testicular cancer occurs only in males. Well, of course it does, but even Hippocrates' contemporaries knew this: males have testes, females have ovaries.

For an amusing perspective on overfitting as an aid to getting your data-mining paper published, read the online article by John Langford [20]. This article was originally written in 2005, but was reposted, on invitation, with modest edits 10 years later, because of its high impact.

10.4 THE HUMAN COMPONENT

While the mechanical aspects of data mining are done with computers, people are critical to the process. The belief that any novice with modern computer hardware, user-friendly data mining software, and a lot of data can be turned loose to discover valuable nuggets of information is grievously erroneous. This belief is, sadly, reinforced by articles such as one by Chris Anderson, then editor-in-chief of Wired, entitled "The End of Theory: The Data Deluge makes the Scientific Method Obsolete" [21].

Silver counters this (preposterous) viewpoint eloquently: "Numbers cannot speak for themselves. We speak for them, by imbuing them with meaning. Data-driven predictions can succeed, and they can fail. It is when we deny our role in the process that the odds of failure rise. Before we demand more of our data, we need to demand more of ourselves" [5].

As in so many other fields, there is no substitute for expertise and knowledge. Knowledgeable humans must be aware of a large variety of techniques, and must tailor the choice of techniques to individual problems. They must prepare the data by transforming it into a structure that is suitable for individual mining methods to work with. Most important, the judicious identification of patterns requires deep domain knowledge, which usually implies a very comprehensive knowledge of the underlying theory and principles of the field—as well as research of the published literature—to identify what is truly novel, interesting, and unexpected.

That is why the term "data science" has now come into vogue: a data scientist is someone who combines technical, analytical, and domain knowledge. The human element is what differentiates data science from data mining. Provost and Fawcett [11] provide the best description of the skillset required to work in data science, and the processes involved. They point out that, unlike the traditional software development cycle, much of the work is necessarily exploratory, with a lot of throwaway prototypes, and is heavily "agile" in demanding the ability to intellectually turn on a dime, figuratively speaking. The importance of findings is checked continually with literature search. Actual "software engineering" comes much later. Among the important skills one looks for in a data-science hire are the ability to work with ill-formed problems, the ability to formulate problems well, design proof-of-concept prototypes rapidly, and design experiments.

10.4.1 The problem of overt or subconscious bias

Also critically important is the freedom from biases, or at least awareness of one's own biases. Silver cites the work of Philip E. Tetlock, then at UC Berkeley, in categorizing analysts into two personality-type extremes: "foxes" and "hedgehogs" (a metaphor borrowed from an Isaiah Berlin essay that says "A fox knows many things, the hedgehog one important thing."). In analyzing the accuracy of political forecasts, Tetlock found that the rigidly ideological "hedgehogs," who were more highly expert in a single field, fared much worse than the intellectually eclectic "foxes," who were not. Tetlock's book, *Expert Political Judgment: How Good Is It? How Do We Know?* [22], is an entertaining and insightful read.

One might think that sciences like biomedicine are less vulnerable to bias, but this is not so: journals tend to publish only positive results, and hence there is a motivation for scientists to get positive results. John Ioannidis points this out in a paper entitled "Why most published research findings are false" [23]. The field of neuroscience appears to be particularly vulnerable, with lots of experiments performed with very small sample sizes, where random effects may appear that are subsequently contradicted by larger-sized studies [24]. (While this chapter is on "big data," the sheer expense of collecting data in the neuroscience field makes it a "small data" field.)

But simply collecting more data may not solve the bias problem. As has been repeatedly pointed out by various authors, our evolution has conditioned us to see patterns in random noise. When you do big-data analytics, you are literally testing thousands to millions of hypotheses simultaneously, and so the chances that one finding will come up "significant" without a Bonferroni correction are close to 100%.

10.5 CONCLUSIONS

Data mining can be rewarding, provided that you are aware of the limitations of the methods employed, of the data sources that you are using, and of your own limitations as a fallible human being when you employ these techniques. The issue of fallibility has been explored in various contexts, and in addition to reading about data mining techniques, you would also benefit from reading about research into thinking. In addition to Nobel Laureate Daniel Kahneman's *Thinking, Fast and Slow* [25] (which explores hidden biases in decision making) is James Adams' timeless classic *Conceptual Blockbusting: A Guide to Better Ideas* [26] to help get you started on the different *kinds* of mental approaches that you need to attack diverse problems.

Using very large volumes of data may help, or it may not. Diversity of data sources—some of which may be free, while others may need to be purchased—is more likely to yield insights if only because findings that turn out significant when applied to two independently created data sources (assuming that your prediction system was trained on one source but tested on the other) are more likely to be genuine rather than false positives.

10.6 ADDITIONAL RESOURCES FOR LEARNING

This section, as I've stated before, combines recommendations for machine learning and data mining because the two are so closely coupled. Unfortunately, no single text offers complete coverage of all the methods that a practitioner would need to employ. Even the authoritative textbooks, such as Hastie, Tibshirani, and Friedman's *Elements of Statistical Learning* [27] (which includes graduate-level math that few clinical computing practitioners can process), omit entire categories of methods.

Books describing the mathematical underpinnings of a science are necessary, of course, because without such underpinnings, mathematicians, statisticians, and computer scientists do not have the necessary intellectual foundation to devise new methods. However, the number of people (in business and the sciences) who need to be able to merely *use* the techniques (while probably willing to take the underlying mathematical principles on trust) vastly exceeds the number of methodological innovators, just as automobile drivers vastly outnumber automotive engineers. Hastie and Tibshirani realized this and, partnering with Gareth James and Daniela Witten, have written a far more accessible version of their work, *An Introduction to Statistical Learning* [28], where the math level rarely exceeds the level required by a high school calculus student. While this book *still* doesn't remedy the coverage issue, it is far more accessible.

For the practitioner, what is important is to know what a particular method does, the circumstances in which you apply it, the assumptions you make implicitly about the data when you use it, and its limitations. Unfortunately, a good book on data mining for the biomedical sciences hasn't yet been written—this may hopefully change. A book which I strongly recommend is *Data Science for Business* by Foster Provost and Tom Fawcett [11]. This book not only describes the techniques involved, but also elaborates on the processes (both individual as well as management-related) that are necessary to operationalize data science as an integral component of research.

A book that complements the above book (with practical exercises in addition) is *Mining of Massive Datasets* by Leskovec, Rajaraman, and Ullman [3]. The authors make this book freely available: it is used as the basis of Stanford's online course on data mining. The book might be a bit too heavy on math for the typical reader of the present book, but if you can skim past the math and stick to the prose, you will be rewarded with excellent first-hand knowledge from authors who are in the forefront of their field.

Andrew Ng's online lecture notes related to his CS229 machine-learning course at Stanford (http://cs229.stanford.edu/) are also a valuable resource for someone more computer-science-oriented. While he includes math, his textual descriptions are clear enough that you can generally skim past the equations. (After typing his name into Google, "CS 229: Machine Learning" appears as the second hit, which means that lots of people use it.) Ng is also one of the co-founders of Coursera, which offers both free and fee-based online courses, and the chief scientist at Baidu, the Beijing-based web-services company whose search engine is the Chinese equivalent of Google. A former student of the class, Alex Holehouse, has put up his own lecture notes for this course at http://www.holehouse.org/mlclass/. These notes serve as a useful summary once you have gone through the lectures themselves.

10.6.1 "Cookbooks"

Given the diversity of software that exists for data mining, "cook book"-type works describing how to use the techniques are necessary—using the author's favorite tool, be it a programming language like R or Python, or special software—and some of these also emphasize principles. The aforementioned book by James et al. [28] has practical examples and exercises in R.

However, the quality here may be spotty. The group at the University of Waikato, for example, created Weka, [29], an open-source data-mining tool, which offers basic functionality. While the software, the first of its kind, represents a valuable service to the community, its user interface is extremely crude—though I have a twinge of conscience criticizing a package that's free. However, the book by the same group [30], intended as a kind of data-mining tutorial and detailed Weka manual, is of indifferent quality, written badly enough to put you off the subject—and you *do* have to pay good money for it.

10.6.2 Learning with free or commercial software

Even if you're not mathematically oriented, you can't even begin to understand machine learning unless you first develop a familiarity with multivariate statistics. The best way to do this, as I've stated in chapter: An Introduction to Clinical Research Concepts, is to play with a user-friendly menu-driven statistics package that is chock-full of examples/ sample data, such as SPSS or MiniTab. You start by exploring each menu option in turn, hitting the Help button, and going to the example/s that accompany each topic: all of the examples tend to be sufficiently realistic that you can readily relate to them.

The program RapidMiner™ [31], available in both free and commercial versions, is a very good first choice that I recommend highly as a tool to experiment with—and if you're hooked, even purchase; for the price of $999, it is a good investment if you need to do mining often, but are not a full-time analyst. Unlike Weka, it is usable immediately after download, the online documentation is reasonably clear, though it is terse and doesn't emphasize the principles underlying the very large number of tools that it provides; for that, you have to go elsewhere.

As I've stated before, R is a free tool, but the learning curve is steep enough that you should stay away from it if you're only a casual user. If you plan to become a data-science professional, however, you may find it indispensable.

BIBLIOGRAPHY

[1] Wikipedia. Information awareness office. Available from: https://en.wikipedia.org/wiki/Information_Awareness_Office, 2015.

[2] Wikipedia. Policy analysis market. Available from: https://en.wikipedia.org/wiki/Policy_Analysis_Market, 2015.

[3] J. Leskovec, A. Rajaraman, J.D. Ullman, Mining of Massive Datasets, Cambridge University Press, Cambridge, UK, (2014).

[4] D. Snow, Adding a 4th v to big data – veracity, http://dsnowondb2.blogspot.com/2012/07/adding-4th-v-to-big-data-veracity.html, 2012.

[5] N. Silver, The Signal and the Noise: Why So Many Predictions Fail – But Some Don't, Penguin Press, New York, NY, (2012).

[6] C. Taylor, Mashable.com Triumph of the nerds: Nate Silver wins in 50 states. Available from: http://mashable.com/2012/11/07/nate-silver-wins/#saskfMlRYaqL, 2012.

[7] A. Rajaraman, More data usually beats better algorithms. Available from: http://anand.typepad.com/datawocky/2008/03/more-data-usual.html, 2008.

[8] M. Torrance, Better algorithms beat more data – and here's why. Available from: http://allthingsd.com/20121128/better-algorithms-beat-more-data-and-heres-why/, 2012.

[9] O. Tawakol, More data beats better algorithms – or does it? Available from: http://allthingsd.com/20120907/more-data-beats-better-algorithms-or-does-it/, 2012.

[10] J. Diamond, Guns, Germs, and Steel, W.W. Norton & Co., New York, NY, (1999).

[11] F. Provost, T. Fawcett, Data Science for Business: What You Need to Know About Data Mining and Data-Analytic Thinking, O'Reilly Media, Sebastopol, CA, (2013).

[12] D. Austin, We recommend a singular value decomposition. Available from: http://www.ams.org/samplings/feature-column/fcarc-svd, 2010.

[13] Wikipedia. K-Nearest Neighbors algorithm. Available from: https://en.wikipedia.org/wiki/K-nearest_neighbors_algorithm, 2015.

[14] M. Goodman, Future Crimes, Doubleday, New York, NY, (2015) 243.

[15] E. Dadachova, R.A. Bryan, X. Huang, T. Moadel, A.D. Schweitzer, P. Aisen, J.D. Nosanchuk, et al. Ionizing radiation changes the electronic properties of melanin and enhances the growth of melanized fungi, PLoS One 2 (5) (2007) e457.

[16] P.M. Tierno Jr., B.A. Hanna, M.B. Davies, Growth of toxic-shock-syndrome strain of *Staphylococcus aureus* after enzymic degradation of 'Rely' tampon component, Lancet 1 (8325) (1983) 615–618.

[17] R.M. Califf, Clinical trials, in: D. Robertson, G.H. Williams (Eds.), Clinical translational science: principles of human research, Academic Press, Los Angeles, CA, 2008.

[18] M. Copeland, The Real Spy World, Sphere Books, London, UK, (1978).

[19] J. Bentley, Programming Pearls, Addison-Wesley Professional, New York, NY, (2000).

[20] J. Langford, G. Piatefsky (Ed.). 11 clever methods of overfitting and how to avoid them, <http://www.kdnuggets.com/2015/01/clever-methods-overfitting-avoid.html>, 2005.

[21] C. Anderson, Wired. The end of theory: the data deluge makes the scientific method obsolete. Available from: http://www.wired.com/2008/06/pb-theory/, 2008.

[22] P.E. Tetlock, Expert Political Judgment: How Good Is It? How Can We Know?, Princeton University Press, Princeton, NJ, (2006).

[23] J.P. Ioannidis, Why most published research findings are false, PLoS Med. 2 (8) (2005) e124.

[24] K.S. Button, J.P. Ioannidis, C. Mokrysz, B.A. Nosek, J. Flint, E.S. Robinson, M.R. Munafo, Power failure: why small sample size undermines the reliability of neuroscience, Nat. Rev. Neurosci. 14 (5) (2013) 365–376.

[25] D. Kahneman, Thinking Fast and Slow, Farrar, Straus, and Giroux, New York, NY, (2011).
[26] J.L. Adams, Conceptual Blockbusting: A Guide to Better Ideas, 4th ed., Basic Books, San Francisco, CA, (2001).
[27] T. Hastie, R. Tibshirani, J. Friedman, Elements of Statistical Learning: Data Mining, Inference, and Prediction, 2nd ed., Springer, Stanford, CA, (2008).
[28] G. James, D. Witten, T. Hastie, R. Tibshirani, An Introduction to Statistical Learning With Applications in R, Springer, New York, NY, (2013).
[29] University of Waikato. Weka 3: data mining sofwware in Java. Available from: http://www.cs.waikato.ac.nz/ml/weka/, 2010.
[30] I.H. Witten, E. Frank, M.A. Hall, Data Mining: Practical Machine Learning Tools and Techniques, 3rd ed., Morgan Kaufmann, San Francisco, CA, (2011).
[31] RapidMiner Corporation. RapidMiner. Available from: https://rapidminer.com/, 2015.

CHAPTER 11

Conclusions: The Learning Health System of the Future

11.1 INTRODUCTION

If one uses the frequency at which the term "informatics" is popping up in the language of clinicians and administrators in the healthcare field who know very little about it, then the future for clinical research informatics promises to be very bright. This is not necessarily a sarcastic comment. Nobody can be expected to know everything: while some may merely be mouthing a buzzword, the sensible ones acknowledge its increasing importance and admit honestly that they need (or have) help in this area. For the reader of this book, this translates into more job opportunities and greater responsibilities.

Until now, however, in most organizations, "research" has always been separate from "practice." "To research" means "to investigate systematically." All through this book, there's been the implicit assumption that research consists of a specific set of processes and actions that are geared toward the systematic investigation of biomedical phenomena, and that such processes are quite different from the demands of routine healthcare. Beginning in 2007, however, the Institute of Medicine (IOM) organized a series of workshops with participation from national experts to investigate the possibility of improving healthcare through a series of workshops that centered on the idea of a "continuously learning health system (LHS)."

The IOM's initial report is available as a set of workshop summaries, downloadable from the National Academies Press website as an 11-volume *Learning Health System Series* [1]. There are several aspects to the basic idea, which has the goal of dramatically improving the quality and efficiency of healthcare delivery by essentially blurring the boundary between research and practice through iterative cycles of improvement. The core ideas are as follows.

1. A change in the culture of the healthcare workforce at all levels (from the lowermost staff to top decision makers) so that every interaction between stakeholders in the system (most commonly between the patient and the healthcare provider) is looked upon as an opportunity to learn and gain experience that is documented.
2. The extensive and indepth use of information technology to collect data that can be processed using big-data analytics to generate testable hypotheses, and perhaps "knowledge" that can be utilized toward the previously mentioned goals.
3. A continual optimization of workflow processes to improve efficiency, again supported maximally by information technology.

Clinical Research Computing. http://dx.doi.org/10.1016/B978-0-12-803130-8.00011-7

Many useful ideas have been generated through the workshops—the reader is encouraged to at least skim through the summaries. My intention here is to go beyond issues already explored in the workshops so that we can embark with clear eyes on the challenges that arise.

11.2 ORIGIN AND INSPIRATION FOR THE LHS PROPOSAL

It is useful to understand the historical context behind the LHS ideas, if only to reinforce George Santayana's dictum that those who fail to learn the lessons of history are condemned to repeat it. The notions behind the LHS borrow from several ideas in the business world. (The idea of cultural or societal change for the better—at least, in the opinion of the person proposing the change—has been around forever, going back to at least Plato, who advocated rule by philosopher-kings.)

- *Continual learning and knowledge sharing*: This idea borrows from the field of "knowledge management" [2].
- *Large-scale data collection*: This borrows from data mining and "big data," whose strengths and limitations have been discussed in a previous chapter.
- *Optimization of workflow processes*: This borrows from the management ideas of business process reengineering [3].

All these ideas have had their share of both successes (some) and failures (many more). Let's consider them individually.

11.2.1 Knowledge management

The idea behind knowledge management (KM) is that the knowledge and skills of an organization's employees are often lost after they retire or move elsewhere, and so they should be captured formally in some form of "knowledge base." KM was a fad in the 1990s—several software products were introduced to support this activity—but now the luster seems to have faded. It has helped in a few areas such as customer technical support (there is no evidence that civilized product documentation at multiple levels such as end-user, administrator, and developer; a simple Wiki; and a continuously updated Answers to Frequently Asked Questions do worse than an expensive tool), but it is far from the panacea it was proclaimed to be.

The numerous reasons for the failures have been categorized by Frost [4]. In addition to the usual failures of technological initiatives due to lack of management leadership and commitment, inadequate budget, or technological naïveté and an inadequate toolset, the following factors stand out.

- *Organizational culture*: In organizations with an authoritarian, punitive, fear-driven, clique-based, or politics-ridden environment, people may be reluctant to volunteer ideas for fear of being put down or simply having their ideas stolen without receiving credit. Others with the knowledge, but not within the "in" circle, will not be asked to contribute. Those with special skills will hoard their knowledge so as to make themselves irreplaceable—or, if they leave, to be sure to take their skills with them so they have an edge in expertise. In other words, many employees who have the knowledge are unwilling or unable to contribute.

- *Lack of incentives*: Even in organizations where the culture is positive, codifying knowledge takes lots of time and intellectual effort. Employees who have other responsibilities may not contribute unless incentives are put in place for doing so. And merely putting the facts down doesn't constitute knowledge transfer: the facts must be described in a lucid fashion, with sufficient overview and background provided. (Explaining a topic clearly to the nonexpert isn't easy: try reading the Wikipedia article on conditional random fields, which almost totally lacks such background, and see if it doesn't make your head spin.) Not many experts communicate well in writing: they must often explain their ideas to collaborators who can write well. Assigning a collaborator requires allocation of resources.
- *Tacit knowledge versus formal knowledge*: Tacit knowledge—activities such as cooking, flying an airplane, or operating complex machinery—is hard to describe completely through the written word. Such activities have to be learned by repeated practice, often through apprenticeship.

11.2.2 Business process reengineering (BPR)

Originally proposed in 1990 by two MIT professors, Michael Hammer and James Champy, the idea behind BPR was to eliminate unnecessary processes rather than improve or automate them [3]. The authors subsequently wrote a best-selling book expanding on this idea [5]. While sensible in principle, BPR seemed to put processes before people and was decried as a rebirth of "Taylorism."

In the late 19th century, Frederick Winslow Taylor pioneered "scientific management" as a means of maximizing worker output: he began with the basic belief that workers were lazy. The inhumanity of this approach was subsequently parodied by Charlie Chaplin in the 1936 movie "Modern Times." (See [6], which includes an amusing clip from the film showing his character, the "tramp," desperately trying to keep up with an assembly line.) Lucille Ball's "Lucy Show" also had a similarly themed episode (available on YouTube) with the brilliant Phil Silvers playing an "efficiency expert."

BPR was subsequently abused by various organizations to justify wholesale downsizing, even when the organization was doing well, merely to pander to Wall Street's idea of efficient operation and increasing stock price. For many organizations, such as Xerox, this approach backfired, as even the good employees got the message and left. Xerox, whose research team in Palo Alto was once responsible for innovations such as the graphical user interface and the mouse, is now a shell of its former self and was recently rated as the 5th worst company in the United States to work for [7].

Thomas Davenport, an early BPR advocate, wrote a critique as early as 1995, "The Fad that Forgot People," [8] stating "Re-engineering didn't start out as a code-word for mindless bloodshed ... [It] treated the people inside companies as if they were just so many bits and bytes, interchangeable parts to be re-engineered ... Once out of the bottle, the genie quickly turned ugly." Hammer and Champy wrote a revised edition of their book in 2006, emphasizing the human component of BPR, but by then, the damage had been done.

11.2.3 Where BPR and KM have succeeded

Both KM and BPR need two prerequisites to succeed.
1. A strong bond of trust between the organization and the employees who contribute the knowledge or help optimize processes
2. An empowerment of all employees down to the lowest level

The core ideas of BPR and KM are not new: only the ideas of employing extensive computerization are. Both ideas, along with rigorous statistical quality control, were pioneered (without the buzzwords) by the great W. Edwards Deming (1900–1993), the New York University professor whose advice on better manufacturing, after being largely ignored in the United States, was adopted by postwar Japanese industry in 1950.

> **At that time, the philosophy of US manufacturing was "planned obsolescence": few items were built to last. The dividends with Deming's approach were seen 20 years later, when Japanese manufacturing outcompeted in the United States in electronics, automobiles, and steel. Japan's highest award for excellence in industrial achievement is called the Deming Prize. Korea and China are now adopting the same principles.**

Deming wrote in a simple and accessible style. Long after he was hailed as a sage in the United States, he wrote a book called *Out of the Crisis* [9], which is a must-read for those who wish to create a learning organization. Deming's humaneness comes through on every page. He points out that a learning organization is incompatible with fragile egos or a rigidly hierarchical organization: managers must be willing to put their pride aside and listen to their underlings, and give their subordinates freedom to contradict them. They must provide autonomy for decision-making, even knowing on occasion that the subordinate may fail. In Deming's analogy, unless one's child is allowed to take a few steps and fall once, or even several times, she/he will never learn to walk—the parent only needs to ensure that the child will not be hurt if she/he does fall.

Taiichi Ohno at Toyota put this dictum into practice, authorizing the lowest worker to stop the assembly line if a problem was perceived, rather than delay (and risk major problems) by merely reporting the problem up the chain of command. Ohno, the inventor of what is now called "lean manufacturing," was revered by his organization's employees with the title *Sensei* (teacher). His story is described in Nayak and Ketteringham's *Breakthroughs!* [10].

Interestingly, the best Japanese organizations were using collective knowledge gathering long before large-scale computerization, using "quality circles," a method pioneered by Deming in Japan, which involves a group of workers who discuss ways to do their work better and to make better products.

11.3 CHALLENGES OF KM/BPR FOR US HEALTHCARE

The question is to what extent the prerequisites of trust, cooperative behavior, and empowerment exist across the US healthcare system today. One of the books in the *Learning Health System* series (*Digital Infrastructure*), considers the theme of "weaving a strong trust fabric," but the exploration turns out to be limited only to security/privacy of patient data, as now mandated by HIPAA. The issue of trust between employer and employees is not considered at all; nor are broader issues of trust between the patient and the healthcare organization. I explore each in turn.

11.3.1 Trust between patients and healthcare organizations

A learning health system requires those being served—here, patients—to participate actively in the process by providing feedback at every stage. (If US auto and electronics manufacturers had listened to consumers who would have liked their purchases to last somewhat longer than the product warranty, they might not have dug themselves into a collective hole in the 1970s.)

Unfortunately, while patients generally trust their doctors—who are increasingly employees rather than self-employed entrepreneurs, as in the past—there is little evidence that this transfers to the healthcare organization. The Press-Gainey patient satisfaction surveys don't ask many of the questions that patients truly care about: the financial issues are notably avoided. Patients often see themselves as treated poorly and regarded as little more than sources of revenue for the healthcare organization. See the litany of complaints on websites like [11].

Economics Nobel laureate Kenneth Arrow, in a highly influential 1963 article [12], pointed out that the relationship between patients and the medical-care industry is highly asymmetrical with respect to information. The traditional economics assumptions that the marketplace is "efficient" because everybody has the necessary data to make informed decisions is not true of medicine: here, the industry knows a lot more about you than you do about them, and therefore individual organizations can, if they wish, abuse that knowledge.

Rosenthal, Lu, and Cram discovered that pricing data on costs of treatment is not freely available: when prices were quoted for hip arthroplasty over the phone, they varied by a factor of more than 8 [13]. More concerning, out of 100 non-top-ranked hospitals surveyed, only 10 freely provided a quote over the phone. If you don't have transparency, you can't have trust.

The US government has belatedly realized that you can't expect healthcare to police itself. The penalties for readmissions after a premature discharge were prompted in part by the "drive-through deliveries" of the 1990s, where newborns and mothers were discharged less than 24 h after delivery [14], often with adverse effects on both.

11.3.2 Trust between employees and employer

Japanese industry's one-time practice of lifetime employment ensured employee loyalty as well as the long-term interemployee bonding consequent to a stable community within the organization. In contrast, the social contract between employers and employees (a transient phenomenon of the 20th century) has unraveled in many US organizations. There is no evidence that US healthcare has a better organizational climate than the rest of US industry or a better record of dealing with their employees. Ostensibly being in the sacred business of saving people's lives doesn't, by itself, make organizations more moral or ethical than those in other fields of enterprise.

Much of US healthcare is for profit. What is more concerning is that a large number of nominally nonprofit organizations in healthcare operate like for-profits in every respect except paying their fair share of taxes: buying up rival hospitals only to shut them down, paying lavish executive salaries, sending bill collectors after indigent patients, and employing antiunion practices. Kaiser Permanente's practice of discharging mothers and newborns 8 h after delivery was exposed by a whistle-blower and led to congressional legislation that prevented discharge before 48 h without the mother's consent [15].

Such "nonprofit" behavior is hardly limited to healthcare, and has given a new meaning to H.L. Mencken's aphorism: "When someone says it's not about the money, it's really about the money." The concerns about healthcare nonprofits are highlighted in Elisabeth Rosenthal's New York Times article [16]. At least the for-profits are honest about what they do.

I therefore have my doubts as to whether KM and BPR will be applied in US healthcare in a more enlightened fashion than they were in the business world. I would like to hope that benevolent organizations will attract the best talent and hardest workers, and use them to competitive advantage, but I am not optimistic. Healthcare is a local business, and when you fall suddenly ill, you tend to go to the nearest (if predatory) hospital rather than the excellent and benevolent one five hundred miles away. Consequently, as in most other service businesses, unfit organizations may survive because their competitors are too distant.

11.3.3 Risks of purely technocratic approaches

Let's assume that, in specific healthcare organizations, the relationships between employees/employer and patients/organization are *not* dysfunctional, and so at least the major barriers to knowledge-gathering and collaboration do not exist. Additional pitfalls can still lurk for the unwary.

Along with BPR, another wholesale overhaul of organizational infrastructure has been in the area of enterprise resource planning (ERP) [17], where software that integrates (or *claims* to integrate) *all* of the functions of an organization—accounting, sales, purchasing, inventory, shipping, product planning, human resources, marketing—is installed. ERP software typically includes "workflow engines," which help an organization set up electronic workflows for various tasks. The cost of ERP systems rivals that of EHRs.

The process of implementing ERP can be far more delicate (ie, high-risk) than a heart–lung–liver–pancreas transplant. While the few successes have been highly publicized, the implementation disasters have been far more numerous.

The ERP organizations' cleverly orchestrated publicity led many other organizations' CEOs to rush out like lemmings in a mass migration over a cliff, to try to duplicate these successes, betting their organizations' future in the process. Several have been spectacular, resulting in significant stock price drops and even bankruptcies [18]. Unfortunately, unlike in ancient times when a defeated general was ordered to fall on his sword, the ones responsible for the debacle may escape unscathed through "golden parachute" clauses in their employment contract: their subordinates end up paying with their livelihoods instead.

Even as late as 2013—more than 2 decades after the technology was introduced and presumably had time to mature—it was estimated that about 29% of implementations fail [19], with 60% having cost and/or schedule overruns [20]. And these analyses of short-term effects don't even consider whether, after all that money and effort, long-term dividends were ultimately demonstrated that offset the initial pain.

The reasons behind these failures—integration challenges and wholesale, dislocating overhaul or replacement of existing processes, even optimal ones, just to conform to the dictates of often inflexible software—have been extensively discussed in the business literature. Even as of this writing, typing "ERP implementation" into Google *still* brings up "failure" very high on the auto-complete list.

In the context of healthcare, integration of diverse healthcare software still remains a challenge that provides software consultants with steady income, and there is no reason to believe that wholesale workflow changes will be any more successful in the healthcare industry.

More important, workflow reengineering appears most effective when done in incremental steps that don't overwhelm the adaptive capacity of humans. Sure, it can be effected with a sense of urgency, but the principle is to go only as fast as your people can, not flog them to go faster. "Big bang" approaches, whether in the world of information technology, businesses as a whole, or national reengineering efforts, such as Mao's Great Leap Forward, have almost invariably failed.

Any new buzzword coined at the federal level can bring out the con men who will trot out shiny (and expensive) new toys that promise to solve your software problems and capture your employees' knowledge or revolutionize your operations. With the history of failed efforts in your mind, you can stay on your guard.

11.3.4 Human dimension

It should be obvious that you can't just order everyone in an organization to think creatively and "learn" from every little experience and expect results. While everyone has equal legal rights, the ability to learn from experience, or even to take advantage of received wisdom that is handed on a platter, is hardly equal. The Bible puts it bluntly in Jesus's admonition to his followers not to "cast pearls before swine." However, the desire to learn continually is a trait that is unevenly distributed, and the employees of a healthcare organization are only a cross-section of the population—not necessarily the best and brightest.

The ability to do research (which is what problem solving, irrespective of scale or complexity, happens to be) takes a particular set of talents, which not all people possess. More important, it takes a degree of passion, where a practical or theoretical problem that you encounter bothers you like a persistent itch that demands scratching. In many great creative efforts (hardly limited to science), creators kept working on something as a hobby or even an obsession, while struggling at poorly paying day jobs and utilizing every spare moment of time toward their passion.

> **Albert Einstein toiled on physics problems while working as a clerk in the Swiss Patent Office. The Indian mathematical prodigy Srinivas Ramanujan worked on mathematics while drawing a salary as an accounting clerk (he would finish his day's allocated work super quickly and, encouraged by a sympathetic boss, would work on math the rest of the day) [21]. The unemployed mathematician Abraham de Moivre supported himself as a chess hustler in a London coffee house.**

Jon Bentley's classic book of essays, *Programming Pearls* [22], is so named by the author by analogy with the process that creates these gems. A bit of grit or food slips into the space between an oyster's shell and mantle. The oyster can't scratch or pick it out. Instead, the animal secretes layer upon layer of mineral around the irritant until a pearl is formed. Bentley states that many great programs begin with an irritant that motivates their creators to find a solution.

11.3.4.1 Creative process

The best book I've read on the discovery process is possibly William Beveridge's 1950 classic, *The Art of Scientific Investigation* [23]. Beveridge begins by stating: "Elaborate apparatus plays an important part in the science of today, but I sometimes wonder if we are not inclined to forget that the most important instrument in research must always be the human mind."

Several factors, in addition to intellectual passion, have led to great discoveries. I describe these next, with examples.

- *Chance/good luck* (serendipity): A case of "dumb luck" led to the invention, in 1856, of the first artificial dye, mauveine, by the precocious 18-year-old William Henry Perkin, who had set out to synthesize quinine (the natural antimalarial drug extracted from cinchona bark) in his home laboratory, using the coal-tar derivative aniline as a starting point. He failed—at that time, little knowledge was available about quinine's molecular structure, and today we know that his experiment would have no chance of succeeding because quinine and aniline have completely unrelated chemical structures. Perkin, however, ended up with an intensely purple substance, which bonded strongly to silk and was not altered by washing or sunlight. Alexander Fleming's chance discovery of penicillin came about when a blue fungus contaminated and killed his bacterial cultures. (The fungus developed while he was away from his lab on vacation. He had locked up his lab so that the cultures stayed unattended.)

• *Natural curiosity*: While Louis Pasteur pointed out that "Chance favors the prepared mind," curiosity is something that you can't prepare for: either you have it or you don't. (All of us are born with it, but in far too many of us, it is gradually extinguished.) The naturally curious have made discoveries by chasing leads generally ignored by others. Edward Jenner's 18th century discovery of vaccination followed his exploration of the "old wives' tale" that milkmaids never got smallpox [24]. In that era, disfiguring smallpox scars were almost the norm. The phrase "milkmaid's complexion" turned out not to be a myth: they contracted cowpox, a bovine virus that caused mild symptoms but conferred significant immunity to the much nastier smallpox.

• *Imagination and intuition*: The subconscious mind often works overtime so that insights come seemingly out of the blue. The most famous of these is August Kekule's discovery of the benzene ring, in 1865, after a daydream of a snake eating its own tail, which triggered the realization that benzene had a ring structure [25].

• *Sheer effort and persistence*: Thomas Edison's quote, "Genius is 10% inspiration and 90% perspiration" is exemplified by Paul Ehrlich, who initiated the field of chemotherapy—inventing and evaluating artificial chemicals in the treatment of disease. Ehrlich tested 605 compounds with disappointing results until compound 606, Arsphenamine, was found to kill spirochetes (bacteria that cause syphilis, which, at that time, was a death sentence if it became chronic).

Following Pasteur's advice to prepare for good fortune, many means of preparation have been advocated. Again, I provide examples.

1. *Write it down, don't rely on your memory*: The brilliant comedian and screenwriter Larry David (creator of "Curb Your Enthusiasm" and cocreator of "Seinfeld") carries a pen and notebook with him wherever he goes in order to capture a fleeting idea lest he forget it later. Nobel laureate Otto Loewi, who discovered that nerve impulses act through the release of substances that mediate their effects, used to keep a notebook by his bedside. When he got his idea in a dream of an experiment to verify his hypothesis, he wrote his ideas down. After waking, however, he couldn't decipher what he had written. The next night, he had the same dream, and this time went straight to his lab to perform the experiment that got him the Nobel Prize.

2. *Saturate your mind and then take a break*: Your subconscious will continue to work. Beveridge's book, cited earlier, quotes Darwin's autobiography. Darwin recalls how his breakthrough idea about the divergence of species occurred to him while traveling: "… I can remember the very spot in the road, whilst in my carriage, when to my joy the solution occurred to me …" [26].

3. *Have a wide variety of interests*: The likelihood of making a novel discovery increases if you have diverse interests: it increases the chance that you will discover solutions to problems by making connections that others don't—and even if you yourself can't do so, you have a greater chance of knowing people who may see such connections after you explain your problems to them.

Spencer Silver at 3M had invented an adhesive which adhered weakly and temporarily to other materials. For 5 years, he tried unsuccessfully to get others at 3M to find uses for it. Arthur Fry, also at 3M, happened to sing church choir on Sundays and would insert pieces of paper as bookmarks into his choir book, which kept falling off or getting misplaced. Having a Eureka moment, Fry realized that the ideal use of the adhesive was to bind paper to paper.

The full story includes Fry's assembling a prototype machine to print the notes in his home basement, the invention by other 3M scientists of a substrate to prevent the adhesive from leaving residues behind on the surface the note was stuck on, and the marketing managers who turned a failing advertising campaign around by giving away free samples to everyone they met. This is narrated superbly in *Breakthroughs!* [10], cited earlier. The authors, analyzing the various factors that caused Post-Its to succeed, also credit 3M's environment and policy of benign nonintervention, in which inventors are never really ordered to stop working on apparently failed ideas.

4. *Have (and block off) the time to think and let your mind wander.* William Henry Davies lamented the pressures and pace of modern life (in 1911) in his brief poem "Leisure," whose first couplet goes, "What is this life if, full of care, we have no time to stand and stare?" [27]. Stress (especially stress that you can't control) lowers the chance that you will be able to create, simply because your mind and body are too preoccupied with more pressing problems.

I'm mentioning this because the collective creative thinking that an LHS requires is hard to do if the healthcare workers on the front lines of patient care are overwhelmed. Burnout is increasingly a problem [28], which dysfunctional and barely usable EHRs can make worse. Doctors who struggle with documentation needs end up spending more time interacting with the monitor and keyboard than with the patient, who often remarks adversely on the lack of eye contact [29]. How many of the clinical signs identified by some of the great physicians of yore would be discovered by today's physician who has only 15 min to devote to a hasty exam plus documentation?

Unfortunately, as discussed in the mobile technologies chapter, voice recognition (and scribes) have been underutilized, often converting doctors into high-priced data entry clerks. Good luck to organizations who expect overworked employees to innovate *en masse*. Even if asked to think creatively, the responses you get may not be worth the few cents' cost of the electronic media they're recorded on. Alvin Toffler, in his book, *Future Shock* [30], cites a psychology experiment by Usdansky and Chapman [31] (in 1960) in which normal subjects asked to perform word-association tests under intense time pressure produced a pattern of errors similar to that seen in schizophrenics—a failure to think abstractly or make complex connections.

I'm not saying that overworked doctors and nurses function like schizophrenics, though when you're overworked, you tend to function on auto pilot and "muscle memory." However, an organization with a reputation for not caring about its employees' workload can't expect its workers to give a damn about putting in extra effort to improve the organization's revenue or reputation, when that effort may yield little or no benefits for them.

11.3.4.2 *Conclusions*

From the foregoing account, you may have surmised that I have my reservations about the LHS proposal. Beautiful dreams have repeatedly run aground on the reefs of cold reality—in this case, the limitations of human nature and the general problems facing US society and industry, of which healthcare is only a microcosm. These problems have been decades in the making, and asking authoritarian healthcare organizations to change the way they've always operated, to get participation from the employee on the lowest rung, may just not happen.

But I guess I'm very cautiously optimistic: if you don't play the Devil's advocate and identify every roadblock that can derail an idea, you haven't done enough to ensure that it will succeed. That's what I hope I've done here.

Enhancements and improvements are likely to be evolutionary, not revolutionary. I hope that benevolent healthcare organizations that genuinely care about both their employees and their patients succeed in using the ideas laid out in the Learning Health Series to gain competitive advantage.

BIBLIOGRAPHY

[1] Institute of Medicine. Learning Health System Series (11 volumes). Available from: http://www.nap.edu/catalog/13301/the-learning-health-system-series, 2011.

[2] Y. Malhotra, Knowledge management for the new world of business. Available from: www.brint.com/whatis.htm, 2001.

[3] M. Hammer, J. Champy, Reengineering Work: Don't Automate, Obliterate, Harvard Business Review, 1990.

[4] A. Frost, A synthesis of knowledge management failure factors. Available from: http://www.knowledge-management-tools.net/A%20Synthesis%20of%20Knowledge%20Management%20Failure%20Factors.pdf, 2014.

[5] M. Hammer, J. Champy, Reengineering the Corporation: A Manifesto for Business Revolution, Revised ed., Harper Business, New York, NY, (2006).

[6] Work and organizations: film clips. Taylorism. Available from: http://www.clipsaboutwork.com/taylorism.html, 2015.

[7] T.C. Frohlich, M.B. Sauter, S. Stebbin, The worst companies to work for. Available from: http://247wallst.com/special-report/2015/06/29/the-worst-companies-to-work-for/, 2015.

[8] T.H. Davenport, The fad that forgot people. Fast Company. Available from: http://www.fastcompany.com/26310/fad-forgot-people, 1995.

[9] W.E. Deming, Out of the Crisis, MIT Press, Cambridge, MA, (1986).

[10] J.M. Ketteringham, P.R. Nayak, Breakthroughs! How the Vision and Drive of Innovators in Sixteen Companies Created Commercial Breakthroughs that Swept the World, Rawson Associates, Arlington, MA, (1986).

[11] Kaiser Permanente: complaints and reviews. Available from: http://kaiser-permanente.pissedconsumer.com/, 2015.

[12] K.J. Arrow, Uncertainty and the welfare economics of medical care, Am. Econ. Rev. 53 (5) (1963) 941–975.

[13] J.A. Rosenthal, X. Lu, P. Cram, Availability of consumer prices from US hospitals for a common surgical procedure, JAMA Intern. Med. 173 (6) (2013) 427–432.

[14] KomoNews.com. "Drive-by deliveries" can lead to serious illness. Available from: http://www.komonews.com/news/archive/4033336.html, 2009.

[15] J. Court, Ending HMO "drive-thru deliveries", The progressive's guide to raising hell: how to win grassroots campaigns, pass ballot box laws, and get the change we voted for. White River Junction, VT: Chelsea Green Publishing; 2010. Excerpt available from: http://www.consumerwatchdog.org/success-story/ending-hmo-drive-thru-deliveries.

[16] E. Rosenthal, Benefits questioned in tax breaks for nonprofit hospitals. The New York Times 12/13/2013. Available from: http://www.nytimes.com/2013/12/17/us/benefits-questioned-in-tax-breaks-for-nonprofit-hospitals.html, 2013.

[17] M. Krigsman, ERP research: compelling advice for the CFO: new research offers important lessons for chief financial officers when buying and implementing enterprise technology. Available from: http://www.zdnet.com/article/2013-erp-research-compelling-advice-for-the-cfo/, 2013.

[18] T. Wailgum, 10 famous ERP disasters, dustups and disappointments. CIO Magazine March 24, 2009. Available from: http://www.cio.com/article/2429865/enterprise-resource-planning/10-famous-erp-disasters--dustups-and-disappointments.html, 2009.

[19] M. Monaghan, Top six ERP implementation failures. March 24, 2009. Available from: http://www.softpanorama.org/Skeptics/IT_skeptic/erp_skeptic.shtml, 2013.

[20] N. Arellano, ERP cost overruns still plague enterprise: report. IT World Canada February 21, 2013. Available from: http://www.itworldcanada.com/article/erp-cost-overruns-still-plague-enterprise-report/47445, 2013.

[21] R. Kanigel, The Man Who Knew Infinity: A Life of the Genius Ramanujan, C Scribners, New York, NY, (1991).

[22] J.L. Bentley, Programming Pearls, 2nd ed., Addison-Wesley, Reading, MA, (1999).

[23] W.I.B. Beveridge, The Art of Scientific Investigation, WW Norton & Co, New York, NY, (1957).

[24] CN. Trueman, Edward Jenner. Available from: http://www.historylearningsite.co.uk/a-history-of-medicine/edward-jenner/, 2015.

[25] Wikipedia. August Kekulé. Available from: https://en.wikipedia.org/wiki/August_Kekul%C3%A9, 2015.

[26] C. Darwin, Autobiography of Charles Darwin. Electronic PDF version, p. 24. Available from: http://www.classicly.com/download-the-autobiography-of-charles-darwin-pdf, 1999.

[27] W.H. Davies, Leisure, Songs of Joy and Others, University of California Libraries, New York, NY, (1911).

[28] B. Monegain, Burnout rampant in healthcare. Available from: http://www.healthcareitnews.com/news/burnout-rampant-healthcare, 2013.

[29] J.A. Linder, J.L. Schnipper, R. Tsurikova, A.J. Melnikas, L.A. Volk, B. Middleton, Barriers to electronic health record use during patient visits, AMIA Ann. Symp. Proc. (2006) 499–503.

[30] A. Toffler, Future Shock (Mass Market Paperback), Bantam Books, New York, NY, (1984).

[31] G. Usdansky, L.J. Chapman, Schizophrenic-like responses in normal subjects under time pressure, J. Ab-norm. Soc. Psychol. 60 (1960) 143–146.

SUBJECT INDEX

Printed in the United States
By Bookmasters